SYSTEMATIC SAFETY

**Safety assessment
of aircraft systems**

E. Lloyd & W. Tye

Civil Aviation Authority London July 1982

ISBN 0 86039 141 8

First published July 1982
Reprinted January 1992
Reprinted March 2002

NOTE:

Since this book was first published nearly ten years ago, developments in technology have resulted in some areas of the text becoming out-dated. In addition, some of the references are no longer valid. Therefore, where advice on modern practices is needed, the Systems Department of the Authority should be contacted at the address below:

Design and Manufacturing Standards Division
Safety Regulation Group
Civil Aviation Authority
Aviation House
Gatwick Airport South
West Sussex RH6 0YR

Printed and distributed by
Documedia, 37 Windsor Street, Cheltenham, England.

CONTENTS

CHAPTER 6
CASCADE AND COMMON-MODE FAILURES

CHAPTER 7
METHODS AND TECHNIQUES OF SAFETY ASSESSMENT

CHAPTER 8
SOME PARTICULAR TECHNIQUES

CHAPTER 9
RECORDS AND REPORTS OF SAFETY ASSESSMENTS

CHAPTER 10
POST CERTIFICATION

CHAPTER 11
CONCLUDING OBSERVATIONS

PREFACE

In the last thirty years there have been immense changes in aircraft systems. Their capability to fulfil tasks, and the variety of uses to which they have been put, have greatly increased. Perhaps as an inevitable consequence, they have become much more complex.

These developments have transformed the task of safety assessment.

During this same period, the aircraft industry has also changed. No longer is the design of major transport aircraft limited to a few big companies. Such aircraft are now produced all over the world, often as collaborative projects with two or more participating countries. This has led to the need for widespread understanding by constructors and airworthiness authorities of the principles and techniques of modern methods of safety assessment.

We, the joint authors, have long wished that a comprehensive description of the basic processes involved in safety assessment were available. As we could not buy one the only alternative was to try to write one! The opportunity to do so occurred following retirement from the pressure of day-to-day work in the Civil Aviation Authority. The final spur to put pen to paper came from working with the Cranfield Institute of Technology to provide short courses on the subject. The papers presented at these courses by many specialists from British and overseas organisations showed clearly that a wealth of information and ideas was in existence simply waiting to be digested and presented in a convenient form.

We make no apologies for having adopted ideas culled from others, but we are certainly grateful to them. Without attempting the almost impossible task of attributing any particular aspects of this book to particular people, we are indebted to many who have permitted us to use their material, and who have made helpful comments. Without such help, and without the benefit of our work with colleagues in the British and European aircraft industries and authorities for many years, this book could not have been written.

In particular we thank C. Caliendi, R. Cherry, R. W. Howard, J. Hollington, M. Laceby, R. M. McManus, M. Morrison and J. C. Wallin, all from the British aircraft or equipment industry; F. W. de Haan of Fokkers, J. H. Michel of Swissair, and P. Toulouse of Aerospatiale; Prof. E. Edwards of Aston University and F. Taylor of Cranfield; and C.A.A. colleagues A. S. Ashmore, G. L. Gunstone, B. L. Perry, P. F. Richards, J. Rye and D. V. Warren.

We are indebted also to the Civil Aviation Authority for sponsoring the publication of this book, and to the Publications Unit of the C.A.A. for their painstaking work. Perhaps we should add that though there could well be some similarity between the views of the C.A.A. and ourselves, any opinions expressed are nevertheless solely attributable to us.

E. Lloyd
W. Tye

THE AUTHORS

E. Lloyd, B.Sc.(Eng), C.Eng., F.I.E.E., M.R.Ae.S.
Prior to joining the Air Registration Board in 1946, Ted Lloyd worked for some thirteen years in the Joseph Lucas Research Laboratory, at Rotax Ltd., and in the Electrical Department of the Royal Aircraft Establishment. In the A.R.B. he was at first in charge of the Electrical Section, later becoming Head of the Systems and Equipment Department. On transfer to the Civil Aviation Authority he became a Director in the Airworthiness Division until his retirement in 1979.

W. Tye, C.B.E., B.Sc.(Eng), C.Eng., D.Sc.(Hon), Hon.F.R.Ae.S.
In the period 1934-44, Walter Tye served with the Fairey Aviation Co. Ltd., the A.R.B., and for two spells at the Royal Aircraft Establishment. His work was in structures, aerodynamics, and airworthiness. On rejoining the A.R.B. in 1944 he became its Chief Technical Officer and later Chief Executive. In 1972 he was appointed a Member of the Civil Aviation Authority and its Controller Safety. Following his retirement in 1974 he became a visiting Professor at Cranfield Institute of Technology until 1981.

The authors have made major contributions to the art and practice of airworthiness; in addition to their close working contact with the British industry, the authors were involved from time to time with international airworthiness matters, including the development of I.C.A.O. Standards and the European Joint Airworthiness Requirements.

The authors' interest in safety assessment stemmed from the airworthiness investigation for which they were responsible in A.R.B. and C.A.A. This subject is one, in particular, on which they write with great authority.

FOREWORD

Documents stating the underlying philosophy or basic approach to achieving safety objectives are only rarely issued by the C.A.A. and do not fall easily into any of the established publications channels. The underlying philosophy of Safety Assessment of Aircraft Systems, and the working out of that philosophy in the methods used to comply with the requirements, is such a case; it is in the hope and the belief that this book will make a contribution to the pursuit of safety that the C.A.A. has undertaken its publication.

In civil aviation, safety assessments are made by the industry. Airworthiness Authorities have developed a broad philosophical approach as a basis for their published requirements; they are also concerned with the adequacy of the individual safety assessments which are made to show compliance with those requirements. Neither of these are matters suitable for issue with the normal "rules".

The system designer's task of making Safety Assessments is not, of course, carried out solely to satisfy the rules of Airworthiness Authorities. The design itself is likely to benefit from lessons learned from a systematic assessment of safety. Therefore it is clearly to everyone's advantage that a widespread understanding of the basis and of the methods and techniques involved is encouraged, particularly today when many aviation projects are multi-national. This book will, it is hoped, aid such an understanding.

Safety assessment of the kind described here is a relatively new and rapidly developing technique. There is room for debate on the most suitable approaches to follow. This is a further reason why a descriptive book appears to be a suitable vehicle for bringing together current ideas.

The development of the philosophy underlying the safety assessment of systems is not unique to civil aviation. A number of industries in which the safety of the public or of the workers involved in the industry is dependent on the adequate design of systems from a safety point of view are also working along these lines and some have developed more or less sophisticated regulations on this basis. It is, therefore, hoped that the principles and some of the methods outlined in this book will be found to be of interest to those tackling the problems of design of safety critical systems in other engineering disciplines.

J. C. Chaplin

1
GENERAL BACKGROUND

THE CHANGING SCENE

In recent years there has been a continuous increase in the application of complex systems to civil aircraft, and the aircraft have become critically dependent on many of these systems for their safety. This trend is likely to accelerate over the next decade.

The origin of these changes goes back to the systems used in transport aircraft designed in the 1940's and 1950's. There were many such systems but they were comparatively simple in design and their satisfactory operation was not always a crucial matter in respect of safety. Also, these earlier systems were mainly self-contained, so that failure of one did not influence the continued safe operation of the others.

The airworthiness requirements of that time, contained, for example in British Civil Airworthiness Requirements (BCAR) and in the U.S.A. Federal Aviation Regulations (FAR), were devised to suit the circumstances. Separate sets of requirements were stated for each type of system and they dealt with the engineering detail intended to secure sufficient reliability. Where the system was such that its failure could result in serious hazard, the degree of redundancy, (e.g. multiplication of the primary systems or provision of emergency systems) was stipulated. Compliance was shown with these requirements by engineering analysis, and where necessary this was supplemented by a 'Failure Modes and Effects Analysis' (FMEA) to determine the effects of single failures. However it was not then the custom to examine fully the effects of combinations of failures, nor did the requirements impose a duty to consider the likely frequency of occurrence of failures.

More recently, the advent of such functions as automatic landing, automatic control of turbine engines and their fuel systems, and high authority auto-stabilisation, and stall identification and protection, had profound effects. These systems were individually much more complex, and there was a considerable increase in the number of interfaces and cross-connections between systems and the effects of the systems on the aircraft. As an example, Fig. 1-1 overleaf shows the numerous interfaces involved in a relatively simple rudder control system provided with a yaw-damper to correct "Dutch-roll" characteristics. It will be seen that in addition to the interconnections between the electrical, instrument, hydraulic, and mechanical systems, there are essential interfaces with the pilot, the power-plant, the structure and the aerodynamics of the aircraft.

The aircraft designer is thus faced not only with the analysis of each system, but also of each function performed by systems acting independently or in concert with other systems or parts of the aircraft. Rapid developments in technology have now opened the way to the achievement of

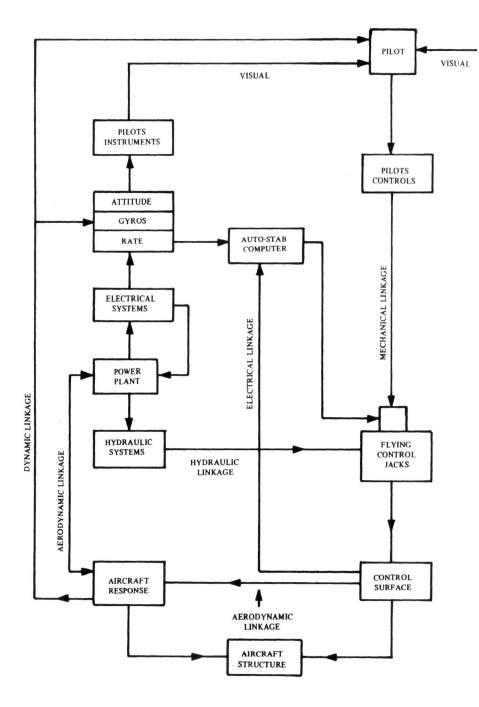

Fig. 1-1   INTERFACES WITH SIMPLE AUTO-STABILISATION
SYSTEM WITH POWERED CONTROLS

2

more and more complex functions together with the integration of signalling or computing components, particularly in the field of flight and power-plant control and of navigation.

Thus the earlier requirements for duplication or triplication have become altogether too crude and uncertain a means of ensuring safety. Had the Airworthiness Authorities continued to issue detailed engineering requirements for each new application before it could be certificated, they would have run a high risk of limiting the development of such systems, and also leading designers into solutions which did not secure optimum safety.

It therefore became necessary to have some basic objective requirements, related to the level of safety, which could be applied to any system or function, and to develop particular detailed requirements as and when experience showed these to be desirable and practicable. It was the advent of the "auto-land" system in the early 1960's which first precipitated this then new approach. It became apparent that by increasing the redundancy of the channels making up the system, increasingly high levels of safety could be secured, and the risk of accident was at least approximately predictable in numerical terms. Thus the need arose to state the acceptable level of risk, for instance in the form of the probability of a fatal accident due to auto-land failure.

Today's requirements, both American and European, follow this new approach of broad objective requirements, backed where necessary by supplementary material relevant to particular types of systems. Modern requirements are discussed briefly below and in more detail in Chapter 4.

THE NEED FOR SAFETY ASSESSMENT
It is evident from the foregoing paragraphs that the safety of a modern aircraft with its assembly of complex systems can only be achieved by a thorough assessment of potential failures, separately and in combination, and of the degree of hazard resulting from such failures. In many, but not in all cases, numerical estimates of the probability of such occurrences will be found desirable or essential.

The requirements have the principle that an inverse relationship should exist between the probability of an occurrence and the degree of hazard inherent in its effect. This is illustrated in Fig. 1-2. The requirements are based on the objective that new designs of large civil transport aircraft should be able to achieve a fatal accident rate of better than one in 10 million hours from all systems causes. (The background to this level is discussed in Chapter 3). Individual features of individual systems can only contribute a small proportion to this target; if one has ten systems with ten critical features in each, this puts the allowable share to each feature at about one in 1,000 million hours, which implies not only very high levels of reliability, which can only be achieved by fail-safe features in some form, but also a very intense scrutiny to obtain reasonable assurance that the target is likely to be achieved. One has to bear in mind that material failures only account for a proportion of the accidents, others being attributable to poor design or maintenance, or errors made by the crew.

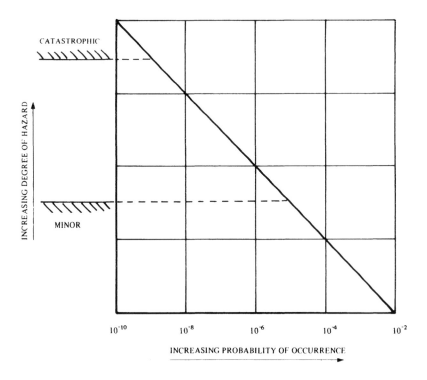

Fig. 1-2   INVERSE RELATIONSHIP OF HAZARD
AND PROBABILITY OF OCCURRENCE

With complex critical systems and functions the designer has not only to consider the effects of single failures, but also the effects of possible multiple failures, particularly if some of the failures are such that they will not be detected by the crew. He also has to consider whether his design is such that it can lead unnecessarily to errors in manufacture or in maintenance or by the crew. In addition the aircraft will be flying in an environment which involves large variations in atmospheric temperature and pressure, and is subject to gusts and other hostile events such as lightning strikes and icing, so that the designer has to consider the performance of each system, both with and without failures, with an extensive range of variables imposed.

In order to maintain and fly an aircraft, detailed instructions regarding checks and emergency procedures are necessary for maintenance engineers and for flight-crews and these must take account of information revealed by a safety assessment. In addition 'allowable' deficiencies which can be permitted to be present before take-off, must be determined and included in the Master Minimum Equipment List (MMEL) (see Chapter 10).

It will be apparent that, in order to achieve and maintain the levels of safety needed, a safety assessment of a system will be necessary particularly when the system or part of it is novel and differs from other systems with a substantial background of technical experience in regard to:–

a. technology,

b. functions,

c. inter-relationship with other systems of the aircraft,

d. relationship between the systems and critical characteristics of the aircraft,

e. complexity.

To determine whether a safety assessment is necessary and to what depth it should be made, the designer will normally conduct a "Preliminary Hazard Analysis" in which he examines the possible hazards arising from a system or function.

He then assesses whether past experience is sufficiently extensive and relevant to justify the system, and which critical features need the more intense evaluation of a safety assessment.

2
ACCIDENTS AND THEIR IMPLICATIONS IN SAFETY ASSESSMENTS

SOME CAUSES OF ACCIDENTS INVOLVING SYSTEMS

Causes of accidents in which systems have played or can play a part can largely be divided into the following categories:–

a. The effects of single and multiple material failures.

b. The lack of adequate performance with or without material failures.

c. Errors in manufacture or maintenance, some of which can be caused by poor design or inadequate procedures.

d. Pilot mismanagement sometimes made possible by poor arrangement of controls or poor presentation of information, or inadequate procedures.

e. The effects of environmental conditions not adequately catered for in the design (e.g. ice, lightning strikes).

f. The behaviour of passengers, ground handlers and cabin crew.

Some of these cases are examined in detail below, and typical examples of accidents are quoted from the "World Airline Accident Summary" (Ref. 1). In order to give an idea of the variety of things that can happen to a critical system three Appendices are attached. The first (Appendix 2-1) lists some serious incidents involving flying control systems over the period 1960–1970. The second (Appendix 2-2) briefly analyses the causes of flying control accidents to piston and propeller-turbine engined aeroplanes over the period 1947–1968, involving about 50 million flying hours, and the third (Appendix 2-3) makes a similar analysis for large turbo-jet aeroplanes over the period 1958–1977, involving about 130 million hours of flying. It will be seen that there has been a substantial improvement in the incidence of accidents caused by equipment failures mainly because of the provision of fail safe systems and because of lessons learned from the past. However, Appendix 2-3 clearly shows that there are an excessive number of accidents on modern jet aeroplanes caused by mismanagement of the control systems by the flight crew.

FAILURES

Types of Failure

A failure occurs when a system or part of a system fails to achieve its specified performance. There are various types of failures which need to be considered including the following:–

a. Single active failures.

b. Passive and undetected ('dormant') failures.

c. Combinations of independent failures.

d. Common-mode failures.

e. Cascade failures.

f. Failures produced by the environment.

7

Single Active Failures

An 'active failure' is one which produces a deterioration in the performance of a system or the aircraft, for example, the failure of an engine to deliver power. Examples of incidents and accidents produced by single failures are most prevalent in the field of flying controls (see Appendices 2-1, 2-2 and 2-3), for example the jamming of controls by foreign objects, the asymmetry of flaps caused by the fracture of linkages, the runaway of electrical or hydraulic actuators, the runaway of automatic controls. There are, however, other sources such as the misbehaviour of propeller reversal and fine pitch controls.

The following are typical examples of accidents:–

4 October 1962   Lockheed Lodestar 18
A malfunction of the electric elevator trim tab unit resulted in aircraft uncontrollability and subsequent structural failure of the right wing.

1 April 1958   Hermes   Meesden Green, Herts.
The accident was caused by the elevator mechanism becoming jammed. This deprived the pilots of control of the aircraft. The jamming was due to the presence of a small extraneous object which entered the control mechanism.

1 December 1969   Heron   St. Thomas
Crashed on final approach, asymmetric flaps due to failure of right flap datum hinge.

Passive and Undetected ('Dormant') Failures

A passive failure produces no immediately observable effect on the performance of a system. However, it can be of a subtle nature depending upon whether or not there is indication of the failure. In a multi-channel system, for example, the failure of one channel of electrical or hydraulic supply should cause no problem provided that the crew is informed of the failure by an appropriate warning or indication. However, if there is no warning of the failure it can remain undetected until either it is picked up by routine maintenance or a further failure occurs.

A typical case of an undetected failure is given below and illustrated in Fig. 2-1.

23 June 1967   BAC 1-11   Blossburg
The probable cause of this accident was the loss of integrity of the empennage pitch control due to a destructive in-flight fire which originated in the airframe plenum chamber and, fuelled by hydraulic fluid, progressed up into the vertical fin. The fire resulted from engine bleed air flowing back through the malfunctioning non-return valve and an open air delivery valve (lack of crew action) through the APU in a reverse direction and exiting into the plenum chamber at temperatures sufficiently high to cause acoustic linings to ignite.

Many modern systems are equipped with monitors to disconnect automatic controls in the event of a malfunction or to give warning to the pilot when dangerous conditions arise. In some cases these monitors can fail in a passive sense so that the failure remains undetected; control of such failures can be exercised by pre-flight checks or checks at other fixed

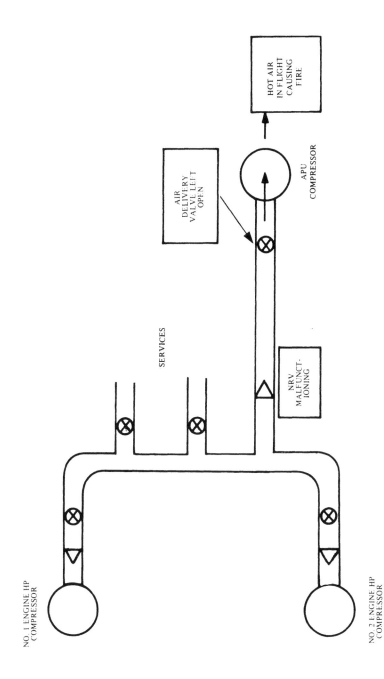

Fig. 2-1 · UNDETECTED FAILURE OF NON-RETURN VALVE
LEADING TO REVERSE FLOW OF HOT AIR AND FIRE

periods. The consideration of such failures plays an important part in safety assessments.

Combinations of Independent Failures

Even if redundancy is provided in the form of emergency systems or duplication, triplication etc., there is always the probability of a multiple failure and such probabilities need careful assessment. There may be combinations of active failures, such as, for example, multiple failures of engines or electrical generators or hydraulic systems. The failures need not all be in the same system; for example, there could be an engine failure on a twin-engined aircraft which combined with the loss of the hydraulic system powered by the other engine would make it impossible to supply sufficient rudder hinge moment to correct for the engine failure.

The hazardous combination of active and undetected failures is usually confined to a particular system but this is not necessarily so. For example, a slow runaway of an auto-pilot may produce an aircraft overspeed condition which is undetected by the overspeed warning because of a passive failure in the latter.

The assessment of combinations of failures needs a very methodical approach combined with considerable knowledge of the aircraft under consideration as well as detailed knowledge of a particular system.

Common-mode Failures

In spite of the provision of redundancy it is possible for the same root cause to affect each part of a system. For example each channel of an electronic system may be affected by electromagnetic or electrostatic interference produced by failures in another system or by atmospheric electrical disturbances, or they may be affected by common errors in manufacture; a local fire or uncontained disc fragments from a bursting turbine engine can produce multiple failures as can untidy electrical wiring causing cross connections.

Fig. 2-2 shows a potential common-mode failure in a duplicated hydraulic system. Here the face seals of two hydraulic systems in a hydraulic jack, (one system acting as a standby to the other) are completed by a single face plate held down by three attachment bolts. The failure of a single bolt can cause the loss of both systems by bending of the plate followed by extrusion of the seals.

Such failures are difficult to predict statistically. The main defences against them are segregation of redundant systems and the use of dissimilar redundancy, as, for example, in the VC10 flying control system where the elevators and ailerons are powered by the main electrical system, these surfaces being split into four each driven by a self-contained electro-hydraulic actuator, while the tailplane and spoilers are powered by the main hydraulic system. The electric system is backed up by a drop-out, ram-air turbine to cover the case of multiple engine failure. (See Fig. 2-3). Chapter 6 deals in detail with common-mode failures.

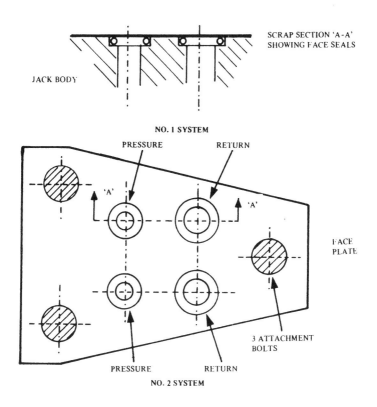

SCRAP SECTION 'A-A'
SHOWING FACE SEALS

JACK BODY

NO. 1 SYSTEM

PRESSURE   RETURN

'A'   'A'

FACE
PLATE

3 ATTACHMENT
BOLTS

PRESSURE   RETURN

NO. 2 SYSTEM

Fig. 2-2   POTENTIAL COMMON-MODE FAILURE IN
HYDRAULIC SYSTEMS USING FACE SEALS

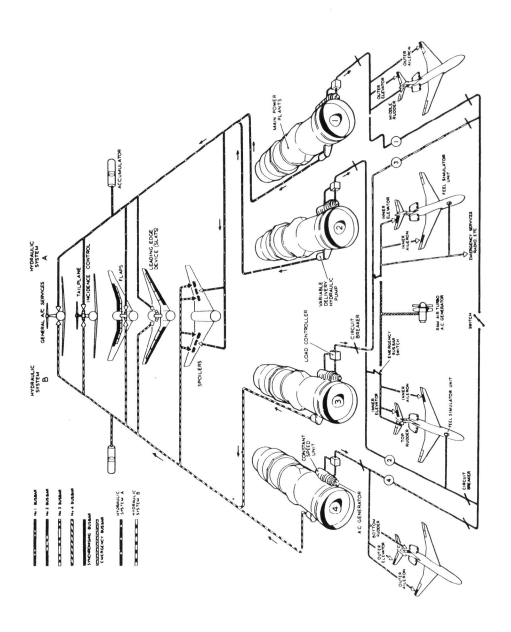

Fig. 2-3   VC10 FLYING CONTROL SYSTEM
(Courtesy of British Aerospace)

12

Cascade Failures

A cascade failure is a particular type of common-mode failure where a single failure, which in itself may not be hazardous, can precipitate a series of other failures. The following are examples of accidents resulting from cascade failures.

5 March 1974   DC10   Paris   (345 fatalities)

The accident was the result of the ejection in flight of the aft cargo door on the LH side: the sudden depressurisation which followed led to the disruption of the floor structure, causing 6 passengers and parts of the aircraft to be ejected, rendering No.2 engine inoperative and impairing the flight controls (tail surfaces) so that it was impossible for the crew to regain control of the aircraft.

The underlying factor in the sequence of events leading to the accident was the incorrect engagement of the door latching mechanism before take-off. The characteristics of the design of the mechanism made it possible for the vent door to be apparently closed and the cargo door apparently locked when in fact the latches were not fully closed and the lock pins were not in place.

This defective closing of the door resulted from a combination of various factors:

Incomplete application of Service Bulletin 52-57.

Incorrect modifications and adjustments which led, in particular, to insufficient protrusion of the lock pins, and to the switching off of the flight deck visual warning light before the door was locked.

The circumstances of the closure of the door during the stop at Orly, and in particular, the absence of any visual inspection through the view port, to verify that the lock pins were effectively engaged although at the time of the accident inspection was rendered difficult by the inadequate diameter of the view port.

Finally, although there was apparent redundancy of the flight control systems, the fact that the pressure relief vents between the cargo compartment and the passenger cabin were inadequate and that all the flight control cables were routed beneath the floor, placed the aircraft in grave danger in the case of any sudden depressurisation causing substantial damage to that part of the structure.

12 June 1975   Boeing 747   Bombay (no fatalities but loss of aircraft)

While aligning for take-off during 180^{0} turn at beginning of runway 27, No. 11 tyre blew out/failed, No. 12 tyre also blew out/failed during take-off run. Following blowing out of tyres on starboard body gear, truck tilted and the wheels and brake assemblies started rubbing the runway surface generating excessive heat which, coupled with hot brakes on 9 and 10 wheels due to overloading and braking action, originated a fire in the starboard body gear wheels.

Due to initial delay in shutting down the engines which hampered the effective fire fighting, coupled with a certain amount of lack of co-ordination and proper deployment of the fire fighting equipment, the fire, originally confined to the starboard body gear, grew into a conflagration and ultimately destroyed the aircraft.

3 December 1976   DC10   Lisbon   (no fatalities)

After take-off from Madrid the flight crew reported failure of Nos. 1 and 3 hydraulic systems and were informed that at least two tyres burst on take-off. Due to bad weather at Madrid the aircraft was diverted to Lisbon where an emergency landing was made on runway 21 with no brakes and No. 2 engine reverser inoperative. The aircraft overran the 3800m runway.

On-site inspection showed:–
Nos. 7 & 8 brake units stuck in mud.
Nos. 2 & 3 tyres burst and Nos. 7 & 8 wheels "nearly missing".
No. 3 anti-skid hydraulic return line broken and the accumulator hanging loose.

Evidence indicated that:–
Heavy compressor stalls occurred on Nos. 1 and 3 engines.
Heavy braking was probably applied on touch down but hydraulic pressure dropped quickly to zero.

(All this was probably initiated by a deflated tyre!)

Failures Produced by the Environment

One has to consider whether systems are specially vulnerable to some environmental conditions particularly if they can cause common-mode failures. An examination of the serious flying control incidents in Appendix 1-1 shows that a significant number have been caused by the accumulation of ice on control equipment such as screw jacks and cables. In one case the source was dripping water from the fresh water system, which froze between the bottom of the fuselage and the controls, to form a block of ice.

Lightning has caused multiple failures of electrical and radio equipment. Serious consideration is needed of its possible effects on complex avionic systems. It is thought that some accidents have been caused by lightning striking an aircraft in the region of a fuel vent and igniting the fuel vapour within fuel tanks.

The systems analyst, therefore, has not only to check the effects of the environment on systems performance but also to consider whether particular environments can cause total failure or serious malfunction of systems and to see that test programmes are adequate to explore these possibilities.

PERFORMANCE

When operating without failures the factors affecting performance may be the variation of system characteristics within their tolerances, the variations of aircraft response and the effects of environmental conditions such as turbulence, windshear, temperature, icing, runway surfaces and many other conditions. In addition, in the case of navigation systems and automatic landing, one has to consider the variations produced by the ground equipment.

In taking account of all of these variables, it may be unreasonable to assume that each variable is at its most disadvantageous limit, so that it is necessary to take account of the statistical distribution of the variables in order to arrive at sensible conclusions.

In the presence of failures a system is likely to have degraded performance so that it may be necessary to reduce the operating envelope of the aircraft to take account of this (e.g. by reducing speed or altitude) or by reducing the demand on the system by shedding loads (e.g. electrical) or by altering the flight plan. Alternatively it may be practical to demonstrate that the combined probability of the failure and that of the other conditions (e.g. gusts) necessary to produce a hazardous situation is acceptably remote.

Accidents produced by lack of system performance have usually been in the field of power-plant or control system performance (e.g. lack of sufficient controllability to recover from flight upsets).

ERRORS

Design Errors

Design errors play a major part in accidents and may in themselves lead to errors by the crew because of poor arrangement of controls or instruments or warning systems, or by maintenance staff because of lack of proper information or the possibility of incorrect assembly of components.

Some sources of design error are:–

a. The failure to take account of all likely environmental conditions (e.g. temperature effects, vibration, icing).
b. The poor segregation of critical systems so that cascade or other multiple failures can occur.
c. Inadequate protection of flammable substances from sources of ignition.
d. The location of essential equipment near sources of contamination (e.g. electrical equipment below toilets and galleys).

Manufacturing Errors

These are basically caused by:–

a. Inadequate control of quality.
b. Insufficient information on drawings.
c. Contamination.
d. Damage.

Manufacturing errors are potentially a prime source of common-mode failures, particularly with electrical and avionic equipment, and one of the objectives of a safety assessment is to identify critical parts so that the manufacturing techniques and controls can be clearly specified. One can scarcely blame an inspector for not carrying out an inspection he was not asked to do, unless the fault is something glaringly obvious.

Maintenance Errors

Typically the maintenance errors that can produce accidents have involved:–

a. Incorrect assembly (e.g. cross-connection, or fitting wrong part or fitting a part in the reverse sense).
b. The carrying forward of defects or the incorrect diagnosis of defects.
c. The leaving of loose objects (e.g. tools) in places where they can cause damage or electrical shorting, etc.
d. Putting wrong fluids into vital systems.
e. The lack of good housekeeping when making modifications and repairs (e.g. the leaving of swarf and loose rivets around).

Some of these can be avoided by design precautions, others by improved procedures or instructions or labelling. They are difficult to forecast, and one has to take considerable account of past experience in order to avoid repetition on new designs. Several examples are given in the Appendices.

Pilot Mismanagement

Pilot mismanagement is a subject sufficiently wide to justify several volumes of its own. This book confines itself to some of those aspects which are related to the design of aircraft systems. Refs. 2 to 7 list the proceedings of various conferences, symposia, etc., which deal in part with this subject.

An examination of a number of accidents involving the erroneous or unwanted operation of controls, levers, switches, etc., shows in the authors' opinion, that although the cause of the accident may have been designated as pilot error, the arrangement of the controls or their method of operation was such that taking account of human fallibility, or "Murphy's Law", the accident was one day bound to happen ("Murphy's Law" has numerous variations, but in essence it states that if it is possible for something to be done wrongly then one day it will be done wrongly).

Table 2-1 summarises an analysis made by the UK Civil Aviation Authority of catastrophic accidents which had their origins to some extent within the aircraft systems, either by failures, by inadequate design, by pilot mismanagement, by combinations of these, or by unknown causes.

TABLE 2-1
CATASTROPHIC ACCIDENTS WITH A SYSTEMS CAUSAL
FACTOR

ORIGIN	No.
Unknown	1
Failure	4
System Design	
a. Incorrect operation by crew after a failure	1
b. Incorrect operation by crew in absence of failure	10
c. System functions incorrectly, but with no components failure or incorrect crew action	4

The following are examples of accidents caused by pilots operating lift or lift-dumping devices prematurely or incorrectly:–

5 July 1970   DC8 Toronto   (109 fatalities)
Preliminary information from the flight recorder indicates that during the approach-to-land the approved procedure for arming the ground spoilers for automatic touch down was not followed. For an undetermined reason the ground spoilers were prematurely deployed, momentarily, resulting in a rapid descent and heavy impact with the runway which caused structural damage to the aircraft.

18 June 1972   Trident   Staines   (118 fatalities)
The immediate causes of the accident were as follows:–
(a)   A failure by the handling pilot to achieve and maintain adequate speed after noise abatement procedures.
(b)   Retraction of the droops at some 60 knots below the proper speed causing the aircraft to enter the stall regime.
(c)   Failure by the crew to monitor the speed error and to observe the movement of the droop lever.
(d)   Failure by the crew to diagnose the reasons for the stick pusher operation and concomitant warnings.
(e)   Operation of the stall recovery override lever.

The underlying causes were these:
(a) The abnormal heart condition of Captain Key leading to lack of concentration and impaired judgement sufficient to account for his toleration of the speed error and (possibly) his retraction of or order to retract the droops in mistake for the flaps.
(b) Some distraction, possibly to be found in the presence of Captain Collins, which caused S/O Ticehurst's attention to wander from his monitoring duties.
(c) Lack of training in the dangers of subtle pilot incapacitation.
(d) Lack of experience in S/O Keighley.
(e) Lack of knowledge in the crew of the possibility or implications of a change of configuration stall.
(f) Lack of knowledge on the part of the crew that a stick-shake and push might be experienced almost simultaneously and of the probable cause of such an event.
(g) Lack of any mechanism to prevent retraction of the droops at too low a speed after flap retraction.

20 November 1974   Boeing 747   Nairobi   (59 fatalities)
The accident was caused by the crew initiating a take-off with the leading edge flaps retracted because the pneumatic system which should have operated them had not been switched on. This resulted in the aircraft becoming airborne in a partially stalled condition which the pilots did not identify in the short time available to them for recovery. Major contributory factors were the lack of warning of a critical condition of leading edge flaps position and the failure of the crew to complete satisfactorily their checklist items.

An accident involving 38 fatalities was probably caused by the pilot switching off the electrical storage battery, his only remaining source of electricity following previous electrical failures, instead of operating a load-shedding switch, thus depriving him of attitude information during a night instrument departure and causing loss of attitude orientation.

As will be seen from Appendix 2-2 numerous accidents have occurred because pilots have attempted to take off with control surface gust locks engaged. This type of accident has been drastically reduced by making it impossible to select take-off power when the gust locks are engaged.

The following is an example of an accident where a relatively minor failure in a system distracted the crew's attention:–

29 December 1972   L-1011   Near Miami   (99 fatalities)
The National Transportation Safety Board determines that the probable cause of this accident was the failure of the flight crew to monitor the flight instruments during the final 4 minutes of flight, and to detect an unexpected descent soon enough to prevent impact with the ground. Preoccupation with a malfunction of the nose landing gear position indicating system distracted the crew's attention from the instruments and allowed the descent to go unnoticed.

The Behaviour of Passengers, Ground Handlers, and Cabin Crew

The behaviour of passengers, ground handlers and cabin crew has to be taken into consideration when trying to predict system failures. For example in the disaster to the DC10 (5 March 1974, Paris) described on page 13 "the underlying factor in the sequence of events leading to the accident was the incorrect engagement of the door latching mechanism" (i.e. by a ground handler). There have been numerous other fatal accidents caused by the incorrect latching of passenger and freight doors.

The following is an example of an accident probably caused unintentionally by a passenger:–

11 July 1973   Boeing 707   Paris   (123 fatalities)

The probable cause of the accident is a fire which appears to have started in the washbasin unit of the aft toilet. It was detected because smoke had entered the adjacent left toilet. The difficulty in locating the fire made the actions of cabin personnel ineffective. The flight crew did not have the facilities to intervene usefully from the cockpit against the spread of fire and invasion of smoke. The lack of visibility in the cockpit prompted the crew to decide on a forced landing. At the time of touch-down the fire was confined to the area of the aft toilets. The occupants of the passenger cabin were poisoned, to varying degrees, by carbon monoxide and other combustion products.

There are other accidents in which injury has been caused by the mishandling of galley and domestic equipment and cinema projectors.

The analyst, therefore, in addition to examining the possibilities of material failures and errors by licensed personnel, has to determine whether the aircraft is vulnerable to predictable actions by other people such as ground handlers, cabin crew and passengers.

INSTALLATION AND SEGREGATION OF SERVICES

Some of the worst accidents caused by cascade failures and other common-mode failures could have been avoided had more attention been given to the design and analysis of the aircraft installations as a whole, taking into account the secondary effects of such things as tyre and wheel bursts, engine explosions or structural failures not in themselves catastrophic.

To make such an analysis only after an aircraft design is completed may lead to extensive and costly re-design at a later stage. It is necessary to make hazard analyses throughout the design process in order to identify, and, as far as is practical, to eliminate the hazards. One technique which has been used is that of "Zonal Analysis" which examines the locations of critical parts of all systems and the effects of failures occurring in other systems in their vicinity. It should also examine the consequences of failure not only of systems but of other airframe parts, engines, etc.

It is necessary to establish firm rules regarding the segregation of critical services so that redundant parts will not receive common damage. Also, when considering the effects of important failures, considerable care is needed to ensure that those systems needed to cope with the effects of the first failure will remain intact. For example, an engine break-up should not cause sufficient damage to the lateral control system to make it impossible for the system to cope with the asymmetry of thrust resulting from the engine failure.

In designing to take account of both normal and unusual failures the 'system architecture' plays an important part. The 'system architecture' is the overall plan of the way the component parts are joined together and to other systems in order to effect the safe and efficient use of redundancy.

POST-CERTIFICATION ASPECTS

When the aircraft enters service, safety assessment does not stop. There needs to be a proper interchange of information between the aircraft constructor and the operators so that:–

a. Modes of failure and critical failure rates which occur in service can be checked against the assessment. If either the particular failure mode or its effect has not been correctly predicted, it is important that the aircraft constructor should know so that he can consider whether the implications are serious.

b. Alterations to critical check and maintenance periods can be substantiated by the assessment.

c. Modifications can be assessed.

d. A sound Master Minimum Equipment List (MMEL) can be maintained and amended according to experience.

SOME GENERAL OBSERVATIONS

It is essential in making safety assessments, even of the most complex systems, that they are comprehensible to all concerned and not just to the analyst. They must assist the designer and management in making decisions. They must make clear what the critical features of each system are and upon which special manufacturing techniques, inspection, testing, crew drills and maintenance practice they are critically dependent.

The purpose of the assessment is not only to convince Airworthiness Authorities that a system is safe, but also to state clearly those aspects, controls, drills, etc., upon which safety depends, so that the standards needed continue to be upheld.

It is important that the assessment becomes part of the total design process, not just something that is done at the end of the development, but something that lives alongside the development with clearly stated objectives so that all concerned are aware of them.

The safety and performance aspects, although important, are not the only ones to be considered. One has to ensure that the design is practical and economical and likely to prove reliable in service. It is relatively easy just to multiply systems to achieve the required level of safety, but this in itself may lead to problems of reliability and spares provisioning in service. In addition, practical operation of the aircraft may demand that it should be able to take off and fly safely, with various defects and shortages present. These factors have to be integrated into the overall design of the system and the aircraft.

References

1 World Airline Accident Summary (A summary of airline accidents involving aeroplanes of more than 5700 kg maximum weight, since 1st January 1946), Civil Aviation Authority (UK).

2 International Air Transport Association (IATA), 20th Technical Conference, "Safety in Flight Operations", Istanbul 1975.

3 Dutch Airline Pilots Association (VNV), Symposium, "Safety and Efficiency: the next 50 years", Den Haag 1979.

4 British Airline Pilots Association (BALPA), Technical Symposium, "Outlook on Safety", London 1972.

5 Airline Pilots Association (ALPA), "Human Factors", Washington D.C. 1977.

6 Royal Aeronautical Society, Symposium, "Flight Deck Environment and Pilot Workload", London 1973.

7 Flight Safety Foundation (USA) (FSF), 32nd International Air Safety Seminar, "New Technology & Aviation Safety", London, 1979.

APPENDIX 2-1
HAZARDOUS FAILURES OF FLYING CONTROL SYSTEMS

1	1960	Elevator control froze up in flight – probably caused by ground crew clearing snow off the aircraft with water.
2	1963	During take off at the moment of rotation stabiliser trim ran away nose down. Eventually slight nose up of attitude was only maintained by large movements of control column.Both pilots' effort required. Caused by faulty brake and inoperative auxiliary brake due to ice formation in brake cavity.
3	1965	Wing drop – control valve binding intermittently.
4	1965	On approach elevator gear stuck in fine – ice on screw jack.
5	1965	Control column forward movement restricted – column gaiter jammed due to pins coming out of guide rails.
6	1965	During and after take off aileron control wheels tended to rotate to right and difficult to hold aircraft level. Bolt jammed between spring retainer and spring housing drain hole – jamming port spring strut in partly extended position.
7	1966	Control restriction while taxying – control column cable loom looped tightly around bearing head on the aileron control wheel boss.
8	1966	Control movement restricted on take off check – bolt on elevator lever fouled servo-motor attachment.
9	1967	Ailerons jammed in neutral – fault in artificial feel unit.
10	1967	Aileron control jammed – freezing of cables.
11	1967	During climb aileron became increasingly difficult to operate. Bolt in leading edge of port inner aileron had become unscrewed and was fouling aileron – defective self locking thread.
12	1968	Aileron boost package stiff – internal fluid leak.
13	1968	On ground check horizontal stabiliser immovable to nose down – fairing plate of vertical stabiliser jammed at top.
14	1968	Rudder jammed on approach. Loose screws fouling control locking lever.
15	1968	Elevator control jammed in mid position. Input lever frozen in 4 inches of standing water – leakage from fresh water system.
16	1968	Ailerons failed during landing. Lever in aileron run fractured at eye end due to excessive torsion during maintenance.
17	1969	Aileron control restriction caused by cracking of auto-pilot servo-motor bracket – causing output lever to fail.
18	1969	Rudder restriction while taxying for take-off. Icing of base of rudder and rudder curtain while standing in freezing rain.
19	1969	Jamming of horizontal stabiliser, caused by omission of locking devices in auto-pilot servo-motor.

20	1969	Fracture of slide valve on flying control jack, caused by faulty heat treatment – occurred on ground.
21	1969	Freezing of control runs in undercarriage bay caused by leakage of water from domestic water system.
22	1969	Failure of leading edge flap jack caused by faulty maintenance leading to erosion of screw thread in jack body.
23	1969	Aileron control rod jammed by electrical cable loom – bonding tag on trim tab rod fouling in aperture. Detected on pre-flight check.
24	1969	Overflow of trays from pantry stowage fouling elevator control cable – detected on ground.
25	1969	Partial seizure of elevator control caused by corrosion of three axis roller assembly at base of control column.
26	1970	Partial seizure of elevator control. Water dripping from water heater in toilet had frozen to a depth of 5 in. – 6 in. on floor of fuselage. (2 incidents).
27	1970	During rotation the crew experienced restriction in the aft movement of the control yoke. Elevator trim was used to become airborne. During the next take off the same problem was encountered. **Cause:** Loose nylon strap fouling elevator controls.
28	1970	During a pre-departure check the engineer found a full quart can of oil resting on the horizontal stabiliser trim controls.
29	1970	Stabiliser jammed in all three modes. Control cable jammed at twin drum and cable guard.
30	1970	In cruise auto-pilot disengaged and aircraft jerked into climb. Stiff nut with sheared off stud had caused intermittent jamming of pulley group.
31	1970	Hammer left in flap housing – seen by passenger on take-off.
32	1970	Aircraft rotated inadvertently below V_1 despite efforts of crew to hold it. Rod end to elevator boost broken and jammed one half of elevator in up position.

APPENDIX 2-2
ANALYSIS OF CAUSES OF FLYING CONTROL ACCIDENTS TO PISTON-ENGINED AND PROPELLER-TURBINE-ENGINED AEROPLANES (WORLD-WIDE 1947–1968)

The following accidents occurred on passenger transport aeroplanes, above 5700 kg AUW. Total flying hours about 50×10^6.

'P' indicates piston engined, and 'T' indicates propeller-turbine engined.

1 CONTROL LOCKS

Type	Year	Accident
P	1947	Failure to remove elevator locking pins before take-off.
P	1947	Controls left locked on take-off.
P	1947	Gust locks moved in flight.
P	1952	Gust locks left engaged plus engine failure.
P	1954	Elevator lock left engaged on take-off.
P	1955	Locks left engaged on take-off, subsequently released.
P	1957	Elevator partially locked. Stalled on landing after bounce.
P	1958	Rudder locked. Ran off runway on take-off.
P	1959	Elevator gust lock not properly secured. No down travel.
P	1963	External gust lock on port elevator not removed. Wheels up landing.
T	1967	Gusts locks badly rigged. Overran runway.
P	1967	Elevator control batten left in. Crashed on take-off.

2 INCORRECT ASSEMBLY

Type	Year	Accident
P	1947	Incorrect assembly of aileron control circuit (reversed).
P	1948	— ditto —
P	1950	— ditto —
P	1954	Failure of elevator trim push/pull rod caused by reversed installation of right elevator trim tab idler.
P	1955	Elevator control incorrectly assembled.
P	1955	Incorrect assembly of elevator tab controls.
T	1958	Reverse operation of elevator spring tab.
P	1961	Failure of aileron primary control caused by incorrect assembly of aileron boost.
P	1963	Improper rigging of rudder bungee system. Overstressing as a result of loss of control.
P	1965	Asymmetric flap condition resulting from failure of right flap caused by cable misalignment.
P	1966	Incorrect rigging of rudder servo-trim causing rudder lock.

23

3 DISCONNECTION

Type	Year	Accident
P	1948	Torque tube of elevator and stabiliser failed from down load – probably due to turbulence.
P	1950	Fatigue failure of aileron control couplings – flutter.
P	1952	Failure of elevator control – poor workmanship.
P	1953	Failure of right aileron tab motor trunnion – subsequent overloading of wing in turbulence.
P	1956	Rudder trim tab adrift – flutter.
T	1957	Failure of bolt in flap unit.
P	1959	Elevator link disconnected from clevis in elevator control horn assembly.
P	1961	Loss of bolt in linkage of elevator boost system.
P	1968	Flap control cable failure.

4 JAMMING

Type	Year	Accident
P	1952	Aileron chain slipped off sprocket.
P	1958	Elevator mechanism jammed by small extraneous object.
P	1958	Control column jammed by direct-view window.
P	1962	Probably jamming of elevator spring tab mechanism.
T	1963	Elevator movement restricted by frozen water in soaked cleaning cloths.
T	1965	Asymmetric flap caused by foreign object jamming flap gear box combined with inadequate indication of flap asymmetry.

5 MALFUNCTION

Type	Year	Accident
P	1962	Malfunction of elevator electric trim tab.
P	1965	Excessive elevator electric trim tab movement.

APPENDIX 2-3
ANALYSIS OF CAUSES OF FLYING CONTROL ACCIDENTS TO LARGE TURBO-JET AEROPLANES (WORLD-WIDE 1958–1977)

The following accidents occurred on passenger transport aeroplanes above 20 000 kg AUW. Crew training accidents are excluded. Total flying hours about 130×10^6.

1 PILOT MISMANAGEMENT

Year	Accident	Fatalities
1962	Inadvertent switching of horizontal stabiliser trim setting on take-off.	15
1963	Loss of control in turbulence due to abnormal position of longitudinal trim.	0
1964	— ditto —	0
1966	Elevator and aileron gust locks not disengaged at take-off.	Not known
1968	Took off with flaps retracted.	3
1970	Spoilers prematurely deployed.	109
1972	Inadvertent deployment of spoilers.	0
1972	Improper retraction of leading edge flaps.	118
1972	Inadvertent selection of spoilers, also engine de-icing not switched on.	72
1973	Inadvertent deployment of spoilers.	2
1973	— ditto —	0
1974	Failure to deploy leading edge flaps.	59
1977	Premature selection of spoilers.	0

2 DISCONNECTION

Year	Accident	Fatalities
1961	Loss of control on final approach — probably a double failure.	72
1974	Foreflap separated.	0
1975	— ditto —	0
1975	— ditto —	0
1977	Elevator control failed.	0

3 JAMMING

Year	Accident	Fatalities
1970	Elevator jammed by foreign object on initial climb.	11

4 MALFUNCTION

Year	Accident	Fatalities
1962	Rudder control malfunction on initial climb.	95
1962	Failure of horizontal stabiliser trim on take off.	130
1972	Failure of auto extension of lift dumpers on landing leading to overrun.	0

3
ACCEPTABLE ACCIDENT RATES

INTRODUCTION

In this book, reference is often made to the 'acceptable probability' of occurrences meaning the probability prescribed in the requirements. This Chapter gives some insight into what is involved in Airworthiness Authorities' activities in establishing these 'acceptable probabilities'.

There are two main aspects. First, there is the matter of determining policy; the level of safety at which to aim. Second, there is the task of choosing the right numerical values in the requirements to secure the intended target. The first part, the policy, is influenced by two considerations. Put very briefly, the Airworthiness Authorities are concerned to ensure that the safety levels are those which the flying public finds acceptable. But they are also concerned that the requirements are not set at such a level as to be impracticable to achieve.

The public demand for safety is a big topic, beyond the scope of this book to discuss in detail. It suffices to say here that the normal passenger is almost certain to opt for a high level, but equally is unlikely to be prepared to pay excessive fares for the benefit of making a high level even higher. As virtually every safety precaution has its associated costs, it follows that the passenger wants a suitable balance to be struck, and the Airworthiness Authorities attempt to do so on the passenger's behalf.

As an impracticably high level of safety is also prohibitively expensive, the two considerations – acceptability and practicability – are closely related. It is perhaps more an opinion than a provable proposition, but it seems to the authors that the level of safety achieved by modern public transport aircraft flown by major air lines is generally acceptable to the public. When in the future, advances in the safety art permit improvements in safety levels without serious addition to cost, then the public may well expect to see safety increased further.

Leaving these more general aspects, the technical task is to establish sensible sorts of numbers in those requirements which are couched in the form of acceptable probabilities of occurrence.

ACCIDENT RATES

The most obvious course to follow is to examine past safety performance in terms of the accident rates achieved. These show what in fact has been practicable, and to an extent past trends give a clue to future expectations.

One of the problems about analysis of accident rates is that the most recent evidence is the most valid, but due to the few accidents which occur in say one or two years, the statistical samples are too small to be reliable. If the period is lengthened for the benefit of studying larger numbers, the earlier evidence may well not be typical of todays circumstances.

Another problem is ensuring that the accident rates studied are relevant to the particular consideration. Published data distinguishes between different classes of service; scheduled passenger, non-scheduled passenger, freight, business, and sporting flying. Distinctions can also be drawn between classes of aircraft. The accident rates between one class and another vary considerably, so a proper choice is necessary. As this book is addressed to safety of systems of the kind used in modern jet transport aeroplanes of types used initially by major airlines, it is the accident rates for this class of aircraft and operation which are most relevant.

FATAL ACCIDENT RATES

As a starting point the fatal accident rate will be examined. This is partly because the fatal accident is clearly an important indicator of lack of safety, but also because fatal accident data are the most reliable. The definition of a fatal accident – one in which one or more persons are killed – leaves no room for varying interpretation of what constitutes such an accident.

Fig. 3-1 shows the fatal accidents per 10^6 flying hours for scheduled passenger services of jet aeroplanes having weights above 5700 kg. They apply world-wide, less USSR and China, and are derived from Ref. 1 updated by Ref. 2. In order to smooth out the scatter of the plotted points, two year periods, instead of annual ones, have been used. Nevertheless, there remains considerable scatter. This is to be expected as the numbers of accidents in the two year periods is fairly small and random variation is at play. However a trend is discernible, and the straight line is the best statistical fit to the plotted points.

The indications are of a steady improvement in rate from about 2 per 10^6 hours in 1960 to about 0·5 per 10^6 hours nowadays. Readers may at first sight think these figures are low, and indeed they are much lower than the more frequently published rates for the world fleets of public transport aircraft. This is illustrated in the upper curve of the Figure labelled "whole fleet", which gives the rates for scheduled services in the world fleets of jets, turbo-props, and piston-engined aircraft combined. This shows considerably higher rates, and the improvement is also less. The reason for the differences is that the safety performance of the mainly older turbo-props and piston-engined aircraft is not nearly so high as that of the modern jets. Moreover, as time has gone on, these older aircraft have been relegated to smaller less well equipped operators, flying from less well equipped airfields, and the accident rate has in fact increased over the years. Fortunately, the proportion of the total hours flown by jets has increased from about 20% in 1960 to over 95% today. Thus, the improving rate of jet aircraft accidents has increasingly pulled down the rate for the whole mixed fleets. By say 1990, the whole fleet may be expected to be jets, and the adverse effect of the poor performance of the older aircraft will then disappear, as indicated by the dotted extrapolation of the upper curve.

The designer of an aircraft needs to look ahead a few years to the time when his design will be in operational service. It is, therefore, necessary to

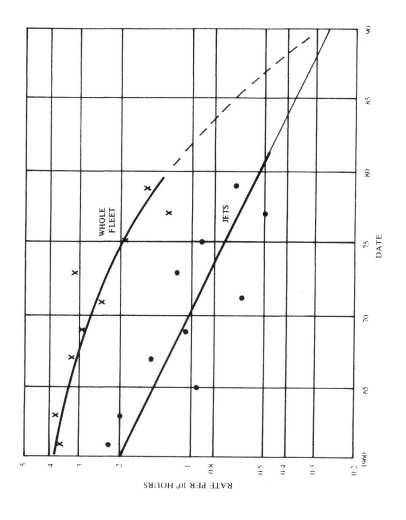

Fig. 3-1   FATAL ACCIDENT RATES

29

make a forecast of what accident rates it might be practicable to achieve in say 1990. If the past trend for jets continues along the straight line shown, then one can expect a rate of under 0·3 per million hours in 1990. However, as safety improves it becomes more difficult to raise it even more, so one cannot take a simple extrapolation for granted.

It helps to consider what has brought about the past trend. Everyone is familiar with the learning curve which shows the improvement in accident rate of a particular type of aircraft during its initial period of flying. The improved safety performance of each type of aircraft during its life makes a contribution to the improved record for the fleet as a whole. However, there is another, and more important learning curve. As each new generation of design appears it benefits from the lessons learned from the previous generation. In parallel with improved intrinsic safety of the aircraft, operational practices are also steadily improving. These together bring about overall improvement in the safety record. It seems reasonable to believe that this process of learning will continue in future. Therefore, while the rate of improvement may not be so high as it has been in the past, it would be wrong to believe that a barrier has been reached.

On this basis, it seems that by the late 80's, while a rate of 0·3 fatal accidents per 10^6 hours might possibly be reached, it would be realistic to assume that 0·4 should be well within grasp.

MEASURES OF AIRCRAFT DAMAGE

Though of less significance than major fatal accidents, those which result in destruction or substantial damage are of importance. An idea of the relative frequency of these classes of accidents can be formed from Tables 3-1 and 3-2, derived from Ref. 2, which distinguishes between aircraft suffering Destruction, Substantial Damage, Minor Damage, No Damage, and Incidents. The figures quoted refer to jet transport aeroplanes, both scheduled and non-scheduled, for the period 1970 to 1978. The figures do not include the categories of Minor Damage or less. Roughly the latter equal in number the Substantial Damage category. An arbitrary distinction is made between Major Fatal accidents in which 10% or more of the passengers are killed, and Minor Fatal in which 9% or less are killed. This distinction does not apply to the freight accidents where the crew alone are affected.

TABLE 3-1
ACCIDENT SEVERITY – PASSENGER JET TRANSPORT

	Aircraft Destroyed		Substantial Damage		Total	
	No.	%	No.	%	No.	%
Major Fatal	84	27	0	0	84	27
Minor Fatal	10	3	3	1	13	4
Non-fatal	18	6	195	63	213	69
Total	112	36	198	64	310	100

TABLE 3-2
ACCIDENT SEVERITY – FREIGHT JET TRANSPORT

	Aircraft Destroyed		Substantial Damage		Total	
	No.	%	No.	%	No.	%
Fatal	10	33	1	3	11	36
Non-fatal	5	17	14	47	19	64
Total	15	50	15	50	30	100

The total numbers of accidents in the freight operations is fairly small and may, therefore, not be a representative picture. Considering the much larger sample of passenger accidents, it will be seen that for every one aircraft destroyed approximately two suffer substantial damage.

As would be expected, major fatalities and aircraft destruction tend to go together. More surprising is the fact that fatalities are rare when substantial damage occurs. The main lessons are:–

a. When the aircraft is destroyed, fatalities occur on about 4 in 5 cases.

b. The ratio Destroyed: Substantial Damage: Minor or no damage, is of the order 1:2:2.

c. The ratio Fatal:Non-fatal is of the order 1:4.

It should be noted that this last conclusion only applies to the average for all accidents. When accidents of particular kinds are considered the ratio Fatal:Non-Fatal is widely variable.

ACCIDENT SEVERITY

Table 3-3 has been derived from an unpublished paper by the British Aircraft Corporation examining accidents to passenger jet transport aircraft for the period 1969/76. Though the samples studied were not exactly the same, they were close enough to result in the same conclusion that the Fatal:Non-fatal ratio was 1:4. The BAC study was such that a more detailed presentation is possible, as in Table 3-3 overleaf.

The wide variation of the tendency for an accident to 'go fatal' is evident. Where accidents involve striking the ground at high speed, e.g. "Striking high ground", there is a likelihood that they will all be fatal. Where the impact is at lower speed, e.g. "Undershoot", the chance of becoming fatal is about 1 in 2. At even lower speeds (the cases marked with a "G") the chance of becoming fatal is small.

In between these extremes there is no golden rule. Much depends on the chain of events which follow the initiation of the accident. If this permits a more or less controlled impact to be made, the risk of fatality is small. If there is loss of control, or serious structural collapse in the air, the risk is obviously high.

TABLE 3-3
TYPES OF ACCIDENT — PASSENGER JET TRANSPORT

	Fatal	Total	Fatal / Total (%)
Predominantly Airworthiness			
Airframe structural failure	1	21	5
Fire (cabin, toilet, etc.)	2	7	29
Landing gear failure/fire (G)	1	20	5
Landing gear mechanism (G)	0	13	0
Engine failure/fire	5	58	9
System failure	7	14	50
Predominantly Operational			
Bird strikes	0	19	0
Weather	6	18	33
Striking high ground	14	14	100
Undershoot	23	45	51
Overshoot/over-run (G)	4	28	14
Running off runway (G)	0	23	0
Heavy landing (G)	0	16	0
Miscellaneous	8	42	19
Total	71	338	21

ALLOCATION OF SHARES OF TOTAL RISK

The total accident rate is, of course, the sum of the rates of accidents arising from a variety of causes. Many years ago, Bo Lundberg of Sweden put forward the idea that the best way to seek to control the overall accident rate would be to allocate shares to the main classes of accident and then to try to control each share. This concept is now often reflected in the practices of Airworthiness Authorities.

In the present context we are concerned with the share which might reasonably be allotted to systems. An analysis of the past record indicates that about 10% of all fatal accidents were attributed to **system failure**. In the period 1970/80 the fatal accident rate was around 0.8×10^{-6} per hour. Hence the portion attributed to systems was of the order 1×10^{-7}, perhaps a little less. It is interesting to note that many of the aircraft concerned were built to meet system requirements intended to achieve a rate of 1×10^{-7}. Intention and reality were quite close.

However it is important to recall the remarks in Chapter 2 about accidents due to improper operation of the systems by the crew in the **absence of system failure**. Table 2-1 of Chapter 2 shows that about half the total catastrophic accidents associated with systems were of this kind. It is likely that in the conventional breakdown of accidents by cause, some generally identified as pilot error may well be ones in which the systems were operated incorrectly. For instance, "flying into high ground" and "undershoot" together account for about half the fatal accidents, and some of these are likely to be system mismanagement ones.

Thus, in the sense that accidents where the crew misuse an otherwise satisfactorily operating system may in part be due to poor system design (e.g. poor warnings or presentation) fatal accidents associated with systems may be nearer 20% than 10% of the total. Looking to the future it is desirable that rates of system accidents should reduce in accordance with the hoped-for general reduction. This would imply that a target of $0 \cdot 3$ to $0 \cdot 4 \times 10^{-7}$ should apply to fatal accidents from system failures. Equally it implies that corresponding improvements in design of systems to reduce the incidence of pilot-induced system accidents are just as necessary.

RATE AND PROBABILITY

The preceding figures have been expressed in terms of accident **rate**. The requirements for systems refer to accident **probability**. There is a distinction to be drawn, though in many cases it has no practical effect. The rate is, in effect, the long term average frequency. Where the cause of an accident is such that there is no reason to believe that its likelihood of occurrence varies from flight to flight, then the probability and rate are numerically the same. An accident rate of 1 per 10^7 hours corresponds to an accident probability of 10^{-7} per hour.

However, one can visualise causes the probability of which is not constant. Fatigue or wear-out failures are examples where the risk of failure increases with hours of use. Thus, theoretically at least, one could have a situation in which, though the overall rate was acceptably low, the probability of failure of old items might be very high.

From the passenger's view point, it is no consolation to know that the average is satisfactory if his particular flight is dangerous. Generally, however, practical steps are taken to avoid this kind of situation, for instance by removing parts subject to fatigue before the onset of high risk. For the most part, in system safety analysis, rate and probability can be taken as one and the same.

There are, however, two special circumstances in which this matter assumes importance. Both are in the category of the acceptance of short term additional risk. One is permitting flights with allowable deficiencies. The other is permitting continued operation pending the incorporation of modifications known to be necessary because of previous accidents or serious troubles.

In both these cases the fact is that the flights are despatched knowing that the airworthiness level is sub-standard. If this were expressed numerically, the probability of accident on the particular flights would be greater than normal. However, because such flights are few relative to the total flying of the aircraft, the effect on the overall accident rate may be imperceptible.

These are matters which Airworthiness Authorities are beginning to consider in a preliminary way. Hitherto, decisions on the acceptability of an allowable deficiency, or the length of the time that an aircraft required to be modified can operate without modification, have not been regulated on a quantitative basis; rather by the 'feel' of the magnitude of the risks.

However, by using safety assessment techniques it would be possible to establish the level of these extra-to-normal risks, and hence, possible to decide on their acceptability by reference to numerical limits. In both cases two considerations would be involved. First, whether the probability of accident on an individual flight was reasonable. Second, whether the cumulative effect of making several flights at above-normal risks would seriously worsen the average long term accident rate.

At first sight, it may seem rather improper to decide on a maximum acceptable probability of accident to apply in normal circumstances, and then to disregard it as a matter of apparent expediency. However, the fact is that aircraft could not fly on a regular basis if they were not permitted to do so with some deficiencies on some occasions. Nor would it be practicable or reasonable to ground an aircraft every time some trouble indicated the need for an improving modification. The only real question is whether to exercise the necessary judgement with or without the benefit of quantitative assessment.

In this connection, it helps to bear in mind two points. First, despite the efforts to achieve uniformly high safety levels in normal operation, there are, in reality, considerable variations above or below average. At any given time, the accident rate of the 'worst' aircraft types is of the order of ten times the rate of the 'best' types. With this unavoidable variation, it is not unreasonable to permit small additional risks to be taken for short periods or for a few flights. The second point is that any particular feature of the aircraft only contributes a small fraction of the total risk. Hence, if one particular feature is temporarily at a much higher risk than normal, the effect on total risk is small.

References
1   Fatal Accident Statistics for Passenger Air Transport Services 1960-1967, CAA Paper 77027, Civil Aviation Authority (UK).
2   World Airline Accident Summary, Civil Aviation Authority (UK).

4
THE REQUIREMENTS

CURRENT CODES OF REQUIREMENTS

Modern system safety requirements are to be found in the principal airworthiness codes, notably the Joint Airworthiness Requirements prepared jointly by several European countries (JAR-25), Ref. 1, and the United States ones in Federal Aviation Regulations (FAR-25), Ref. 2. Both these codes are built on the same underlying concepts, and at the time of writing this book (mid-1980), are closely similar in most particulars. British Civil Airworthiness Requirements (Section D), Ref. 3, includes similar requirements, but by virtue of the United Kingdom decision to accept JAR-25 in lieu of Section D of BCAR, the latter document is becoming of historic interest only. Likewise, the code developed by France and Britain for certification of Concorde also includes a version of modern system requirements, but will become a unique but dated statement.

For purposes of the following description, reference will be made to JAR-25, but for the most part it is equally applicable to FAR-25.

BASIS OF MODERN SYSTEM REQUIREMENTS

JAR-25 treats systems **as a whole**. It requires that the degree of severity of the 'Effect' of all conceivable occurrences be taken into account. Effects are the end results – accidents, incidents, or minor deficiencies. Occurrences are the initial causes – failures of components, separately or in combination, human errors, and external circumstances. The interaction of all these within a system, and between one system and another, is required to be considered.

The broad intention is that Effects of a catastrophic kind should virtually never occur in the fleet life of a type of aircraft. Where the Effects are less hazardous, they are permitted to occur more frequently. The general aim is that the acceptable probability of an Effect should be inversely proportional to its severity.

Table 4-1 is similar to one published in JAR-25, and it neatly summarises the whole structure of the requirements.

In the bottom line of the Table, the Categories of Effects are distinguished relative to their degree of hazard – Minor, Major, Hazardous, and Catastrophic. At the top of the Table, an indication is given of the severity of effects each Category is intended to encompass.

It should be noted that a Catastrophic accident is not quite the same as the conventional definition of fatal accident. The latter covers accidents in which as few as one person is killed, (e.g. a fatality in a flight in severe turbulence). The Catastrophe is in effect a 'multi-fatality' accident. It is true that the definition also includes loss of aircraft, which might sometimes be non-fatal, but in practice it usually involves fatalities.

TABLE 4-1
RELATIONSHIP BETWEEN PROBABILITY AND SEVERITY OF EFFECTS

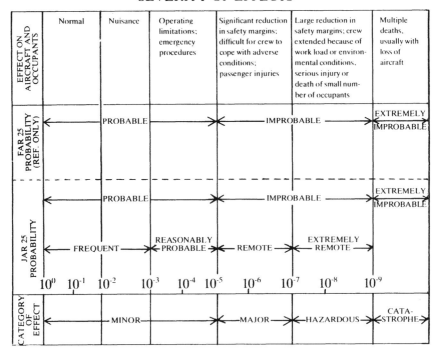

The Hazardous Effect is intended to cover serious incidents in which the risk of catastrophe is potentially high (might be worse than 1 in 100) and also accidents in which a small number of persons may be seriously injured or, exceptionally, killed.

The Major Effect is intended to cover incidents in which the crew could reasonably be expected to continue the flight to make a safe landing, notwithstanding the difficulties of so doing.

A Minor Effect is one in which the airworthiness and/or crew work load are only slightly affected. Some of these might be regarded as reportable occurrences.

For each category of Effect, the centre lines in the Table give indications of the acceptable frequency of each Effect. The fact that the Table quotes numerical probabilities does not mean that all safety assessments must be subject to numerical assessment. Much depends on the simplicity or complexity of the particular system, and in other Chapters of this book the circumstances in which numerical assessment is helpful will be made clear.

LEVEL OF SAFETY IMPLIED BY THE REQUIREMENTS

Some idea of the level of safety implied by the numbers quoted in the Table can be formed by considering an example. A single aircraft might fly a total of 5×10^4 hours, and a large fleet of 200 aircraft might accumulate a fleet total of 10^7 hours. Thus:–

a. An Extremely Improbable Effect (at worst 10^{-9}) would be unlikely to arise in the whole fleet life.

b. An Extremely Remote Effect (at worst 10^{-7}) might arise once in the whole fleet life.

c. A Remote Effect (at worst 10^{-5}) might arise once in an aircraft life, and would arise several times in the whole fleet life.

d. A Reasonably Probable Effect (between 10^{-5} and 10^{-3}) could arise several times in the aircraft life.

THE MAIN REQUIREMENTS

With the foregoing description of the intentions and general structure of the requirements in mind, we can turn next to particular statements in the requirements. Both JAR-25 and FAR-25 state in identical terms the key-stone requirement:–

> The aeroplane systems and associated components, considered separately and in relation to other systems, must be designed so that . . . .
>> the occurrence of any failure condition which would prevent the continued safe flight and landing of the aeroplane is extremely improbable . . . .

In terms of the words and numbers used in the Table, this means that the maximum probability of Catastrophe arising from any individual failure condition must not exceed 10^{-9} per hour.

As was mentioned earlier, the broad intention is that system failures resulting in Catastrophic Effects should virtually never occur in the whole fleet life. If one assumes that as many as 100 individual Catastrophic Effects emerged from the safety assessment of all the systems in an aircraft type, the total maximum risk implied by the requirements would be $100 \times 10^{-9} = 10^{-7}$. This would mean that with a fleet of 100 aircraft of a type, each flying 3,000 hours per annum, one or other of the various Catastrophic Effects might be expected to turn up once in 30 odd years, which is close to the concept of "virtually never".

For Effects of less severity, the requirements permit greater probability of occurrence. For instance a Major Effect is acceptable provided its probability is remote, i.e., 10^{-5} or less. In their simplest form the requirements reduce to the relationships shown in Table 4-2.

TABLE 4-2
RELATIONSHIP BETWEEN EFFECT AND PROBABILITY

Effect	Permissible Maximum Probability (per hour)	
Catastrophic	Extremely Improbable	10^{-9}
Hazardous	Extremely Remote	10^{-7}
Major	Remote	10^{-5}

Minor Effects have been omitted from the above Table, as they are not usually of concern in certification. If they are frequent, for instance of the order 1 in 1000 hours, they may well be operationally and commercially unacceptable.

The need to limit the probability of Catastrophic Effects is self-evident. The reason for limiting less serious Effects deserves some explanation. Consider Hazardous Effects. This requirement serves two purposes. First, the Hazardous Effect includes serious incidents from which the aircraft and its occupants emerge unharmed though the risk of the incident developing into a serious accident is high (possibly worse than 1/100) and is only avoided by exceptional skill on the part of the crew or because conditions are favourable. Thus, there is always the possibility that a less skilled crew performance or more adverse conditions would convert a harmless result into a dangerous one. By limiting the frequency of such incidents, one is indirectly limiting the risk of the occasional serious accident which arises from these added, and somewhat unpredictable, factors. Second, the definition of Hazardous Effects includes accidents which cause damage to the aircraft, and small numbers of injuries or deaths. These accidents themselves justify imposing some limit on their frequency. The Hazardous Effect requirement is thus useful in its own right and also serves as a back-up to support the Catastrophic Effect requirement.

THE MEANING OF 'PER HOUR'

JAR-25 stipulates that:–

> The probabilities should be established as the risk per hour in a flight where the duration is equal to the expected mean flight time for the aeroplane.

The reason for the reference to the flight duration is not immediately obvious, but it does in fact have an important bearing on the application of the requirement.

It helps to consider the influence of time of flight on probability of occurrence. This is dealt with in Chapter 5, but the conclusions drawn there can be anticipated.

Some failures are associated with a particular event in the flight, e.g. failure of flaps to retract, and have nothing to do with the duration of the flight. For such failures, the appropriate measure of exposure to risk is the number of occasions the system operates during the flight.

Other failures, in a system which runs continuously, are associated with the time in hours for which the system runs. This Category includes flying controls which operate throughout the flight, or de-icing equipment which operates for a portion of some flights. The original failure data, on which the assessment is based, are likely to be in the form "per hour" or "per use", as is the most appropriate to the case.

For failures where the probability is p per hour, and the risk is present for say t out of the total duration of T hours, then the total risk in the flight is pt, and the average risk per hour is pt/T. If the risk is present throughout the flight, the average risk is clearly simply p.

However, it is often the case that failure conditions are the result of a combination of individual failures. The arithmetic of this is discussed in Chapter 5, but as one example we can take the total risk in the flight as $(pT)^2$. In this case, the average risk per hour is then p^2T. The longer the flight the greater the hourly risk becomes.

Then there are the failures associated with the numbers of uses in the flight. If the probability of failure is p', and the system is operated n times in the flight, the total risk is $p' \times n$, and the average risk per hour is np'/T. Thus, in this case, the longer the flight the smaller the hourly risk becomes.

It is for these reasons that the choice of a representative duration T to suit the typical flight time of the type of aircraft is of some significance.

SINGLE FAILURES

Usually system design is such that Catastrophic or Hazardous Effects are the consequence of a combination of two or more failures. However it is not possible to avoid entirely single failures which can precipitate a serious Effect. These are usually of the 'common-mode' variety, described in Chapter 6. It is not practical to calculate the probability of occurrence of such rare events, as low as 10^{-7} or 10^{-9}, when they arise from a single failure. For this reason, the requirements make special reference to single failures which could lead to Catastrophic or Hazardous Effects.

In essence, they require a background of actual service experience, relevant to the type of component, supported by detailed engineering evaluation and, where applicable, testing of the component. In short, where numerical estimates of probability would be meaningless, recourse must be had to traditional engineering processes to justify confidence that the exceedingly high reliability needed will be achieved.

FAILURES, ERRORS AND EVENTS

In principle, application of the requirements leads to taking account of combinations of one or more system failures, together with errors by flight crew or maintenance personnel, and conditions external to the aircraft, such as turbulence, lightning, and so forth. The probability of failures and of encountering extremes of external conditions, are reasonably amenable to numerical evaluation. Human errors are not so readily quantified, and conservative assumptions may well be necessary. The codes give what advice they can at the present state of the art.

RELATIONSHIP BETWEEN REQUIREMENTS AND ACCIDENT EVIDENCE

The object of Chapter 3 was to analyse accident rates to provide a basis for choosing suitable risk levels in the requirements and it emerged that the fatal accident rate for systems failures had been of the order 1 in 10^7 hours. It was also shown that fatal accidents and what the requirements define as Catastrophic Effects were numerically much the same. As mentioned earlier in this Chapter, the system requirements are intended to be consistent with a

Catastrophic rate of 10^{-7} so the requirement and the actual achieved levels are close together.

As regards Hazardous Effects, there is no direct correspondence between requirements and accident evidence. The Hazardous Effect is intended to cover potentially risky incidents, as well as minor accidents. The count of accidents, however, only includes the occasional notably serious incident and omits the many apparently insignificant ones. The latter may well go unreported. Hence the ratio of 1:4 for fatal:non-fatal accidents quoted in Chapter 3 has little bearing on the level selected in requirements for the Hazardous Effect Event.

General accident data can also sometimes be a help in the process of showing compliance with requirements. For instance a safety assessment might show that a failure condition could lead to a wheels-up landing. This would clearly be categorised as a Hazardous Effect. But some wheels-up landings could be expected to have more serious consequences, and the question would be whether the Catastrophic Effects requirements were met. General accident evidence would provide a fair indication of the ratio of fatal:non-fatal wheels-up landings, and this would be useful guidance in reaching a conclusion.

References

1 Joint Airworthiness Requirements, JAR-25, published for the Airworthiness Authorities Steering Committee by the Civil Aviation Authority (UK), including Amendments up to Amendment 6, effective 19th December 1979.
2 Federal Aviation Regulations, FAR-25, USA.
3 British Civil Airworthiness Requirements, Section D, Civil Aviation Authority (UK), including Revisions up to 1st October 1976.

5
PROBABILITY METHODS

INTRODUCTION

Purpose of Chapter

Aircraft designers have used probability for a very long time. The designer who first decided to duplicate essential bracing wires in World War I combat aircraft to reduce the risk of the wings being shot away applied an intuitive probability judgment. The difference between then and now is that an instinctive 'feel' of probability has been replaced by quantitative assessment.

For purposes of aircraft design and safety assessment, no extensive knowledge of probability methods is needed. Properly used, simple methods provide a useful tool to aid the processes of design. But though the rules are simple, they should be well understood, if pitfalls are to be avoided.

This Chapter introduces the subject to designers or assessment engineers unfamiliar with the topic.

Most introductions to probability make reference to games of chance. This Chapter is no exception, as tossing coins or throwing dice provide examples of probability methods using numbers which are small enough to be easily comprehended.

Symbols Used

P — Probability of an event occurring.
Q — Probability of an event not occurring.
p — Probability per unit time, usually per hour.
T — Fixed period of time.
t — Elapsed time.
NOTE: Text books on statistics may use alternative symbols to those quoted above.

The Probability Scale

There is a one in two chance that a tossed coin will fall head up. On the conventional probability scale, this would be written,

$$P = 0.5.$$

The chance of drawing the ace of hearts from a pack of playing cards is 1 in 52, and this would be written,

$$P \simeq 0.02.$$

Thus, the probability scale runs from zero to unity. Near zero means near certainty that an event will not happen. Near unity means near certainty that it will happen.

On this scale of probability, if the probability of an event occurring is P, then the probability of it not occurring is $(1 - P)$.

Aircraft Systems – Typical Numbers

Many of the items used in aircraft systems have a probability of failure of once in around 1,000 hours. On certain assumptions, this means a probability of failing per hour of use of 1/1,000, which would be written,

$$p = 10^{-3} \text{ per hour.}$$

Major accidents occur about once in 1 million hours. Accidents associated with some particular cause may be as rare as once in 100 million hours. These probabilities would be written respectively as,

$$p = 10^{-6} \text{ per hour and } p = 10^{-8} \text{ per hour.}$$

COMBINATIONS OF OCCURRENCES

Simple Combinations

With the large numbers of component parts used in aircraft systems, each of which is moderately likely to fail, we are often concerned with the probability of combinations of failures occurring coincidentally.

Beginning with a coin-tossing example, if a coin is tossed twice, there are 3 possible results, 2 heads, 2 tails, or one of each. There is a temptation to conclude that the chance of 2 heads is 1 in 3 (P = 0·33). But this is wrong.

There are in fact 4 possible outcomes:–

First Throw	Second Throw
H	H
H	T
T	H
T	T

So the true probability of tossing 2 heads is 1 in 4 (P = 0·25) and the probability of tossing one head and one tail is 2 in 4 (P = 0·5).

In short, we have to be very careful in calculating probabilities of combined occurrences to be sure we have properly counted the various possible combinations.

The calculation of combinations of failures is much the same as the coin tossing example. There is one important difference. The result of tossing a coin once cannot possibly influence the result of the second throw. But in an aircraft system, one failure may not be wholly independent of another. The designer strives to achieve independence, but cannot always succeed. However for immediate purposes we will take the 'ideal' case in which failures are completely independent of one another.

Taking the simplest case of two items, A and B, with failure probabilities respectively P_A and P_B. The four possible combinations are,

A fails,   B fails;
A fails,   B does not fail;
A does not fail, B fails;
A does not fail, B does not fail.

The probabilities of these four combinations are,

$$P_A \times P_B$$
$$P_A \times (1 - P_B)$$
$$(1 - P_A) \times P_B$$
$$(1 - P_A) \times (1 - P_B).$$

In aircraft system applications probabilities such as P_A and P_B are often of the order of 10^{-3} or less. It follows that the probability of no failure is very close to unity. In other words, with virtually no loss of accuracy, figures such as $(1 - P_A)$ can be taken as 1.

This simplifies the arithmetic, and we can write,

Probability of A **and** B failing $= P_A \times P_B$.
Probability of A **or** B failing $= P_A + P_B$.

This illustrates an important result which frequently appears in safety assessments, namely that the probability of,

A **and** B occurring is the product $P_A \times P_B$.
A **or** B occurring is the sum $P_A + P_B$.

If the two items are identical so that $P_A = P_B = P$, the probabilities are, of,

A double failure P^2.
A single failure $2P$.

In fact, the above is generally applicable for any number of items. The probability of coincidental failure of all the items is,

$$P_A \times P_B \times P_C \ldots \text{ etc., or } P^n \text{ for n identical items;}$$

and the probability of single failure is,

$$P_A + P_B + P_C \ldots \text{ etc., or } nP \text{ for n identical items.}$$

Application of the Above 'Rules'

The practical relevance of the above 'rules' is as follows:–
A system or a sub-system might comprise a number of items linked together in series, to form a 'channel' as illustrated below.

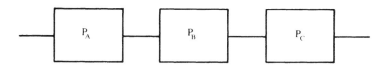

It is evident that with this arrangement, if any one of the items fails, the channel as a whole fails. So here we are concerned with the sum of the probabilities. The probability of channel failure is,

$$P = P_A + P_B + P_C \ldots$$

A complete system is likely to consist of two or more channels arranged in parallel, as illustrated below. The system performs its function if any one channel performs properly.

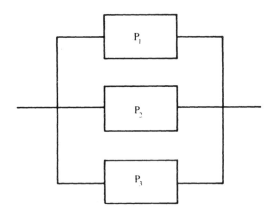

With this arrangement, the channels would all have to fail to cause complete failure of the system. So here we are concerned with the product of the probabilities. The probability of total failure is,

$$P = P_1 \times P_2 \times P_3 \ldots$$

As a numerical example, suppose that a system was made up of identical channels in parallel, each with a failure probability $P = 10^{-3}$. The failure probabilities are,

No. of Channels	Failure of Single Channel	Failure of All Channels
1	$P = \quad 10^{-3}$	$P = 10^{-3}$
2	$2P = 2 \times 10^{-3}$	$P^2 = 10^{-6}$
3	$3P = 3 \times 10^{-3}$	$P^3 = 10^{-9}$
4	$4P = 4 \times 10^{-3}$	$P^4 = 10^{-12}$

Note that multiplexing the channels is a very powerful way of reducing the risk of total failure. But note also that the more the number of channels, the more the number of single failures and hence the worse the serviceability.

More About Combinations of Failures

In a complex system it may be necessary to know the likelihood of partial failures, when some but not all the channels fail. Consider for instance a system of 3 identical generators in parallel. When 2 of the three generators fail, the system continues to operate, but at much reduced capacity, so the probability of this circumstance is important.

The possible failure combinations are,

Triple failure	A and B and C	P^3.
Double failures	A and B $\quad P^2$)	
	A and C $\quad P^2$) totalling $3P^2$.	
	B and C $\quad P^2$)	
Single failures	A or B or C	3P.

As before, the complete failure and the single failure cases are respectively P^3 and 3P. However, the partial failure (i.e. double failure) works out at $3P^2$. Note that this last possibility is $3P^2$ and not just P^2, because there are 3 ways in which double failure could occur.

It is simple to extend this line of reasoning to any number of items. The results for a number of cases are as follows:–

No. of Channels	No. of Failures			
	1	2	3	4
1	P	—	—	—
2	2P	P^2	—	—
3	3P	$3P^2$	P^3	—
4	4P	$6P^2$	$4P^3$	P^4

It will be seen that when the number of channels is large, the risk of multiple failures is to some extent increased because of the number of possible combinations of multiple failure.

The Majority Vote Technique

In some systems, such as the triplex auto-land system, there may be no means of informing the pilot directly which particular channel has failed. The system itself can detect that one channel disagrees with the other two, and it can be arranged for the system automatically to switch off the one in disagreement, on the assumption that this is the one likely to be at fault. The system then continues to operate satisfactorily on the two channels. However, should a second failure then occur, the system has no means of telling which is behaving incorrectly, so total failure ensues. Thus the total failure risk is in this case the probability of two of the three channels failing, i.e. $3P^2$.

Common-mode Failures

All the above has assumed the 'ideal' case where one failure has no influence on the likelihood of other failures. In practice there are important variations from this ideal. The first is the common-part failure. For example, three totally independent flying control systems may merge together in a common part, the pilot's control column. A failure of the common part causes total system failure.

The second is the common-cause, or common-mode, failure. For instance, a fire in a compartment might destroy all the channels of a system running through the compartment. Likewise contaminated hydraulic fluid could cause all the channels of a hydraulic system to fail.

The third is sometimes called the cascade failure, where the result of a single failure is to overload the remaining channels, thereby increasing the probability of their failure.

These common-mode failures and cascade failures can sometimes be obviated by design. However, where they are present, it is vital to take them into account in safety assessment.

Consider as an example a system with three channels, each with a probability of failure of 10^{-3}. The probability of coincidental failure of all three channels is,

$$(10^{-3})^3 = 10^{-9}$$

which is a satisfactorily high level of safety.

However, suppose that there were a common-mode failure assumed to affect all three channels and having a probability of 10^{-7}. The result would be a failure risk of,

$$10^{-7} + 10^{-9} = 1.01 \times 10^{-7}$$

which might be unsatisfactory.

The important point to note is that even when a common-mode failure has a very low probability of occurrence, it can nevertheless totally dominate the overall risk, and can largely destroy the safety secured by the multiplicated channels.

Much could be said about design to avoid common-mode failures, but this is not the purpose of this Chapter. The arithmetic is straightforward, as the above example shows, but for the problem of determining what probability to attach to the rare common-mode failure. The failure probabilities of the individual channels of a multiplex system (often of the order 10^{-3} or 10^{-4}) can be established with reasonable confidence from past experience or specific test. But there is usually lack of sufficient evidence to fix with confidence the failure probability of the extremely remote (e.g. 10^{-7} or less) common-mode failure.

If, as in the numerical example given above, the common-mode failure dominates the total risk, it may be found necessary to provide a back-up or standby of some kind. If the back-up system has fundamentally different design characteristics as compared with the main system, then the same common-mode failure is unlikely to cause both to fail. The back-up system often needs comparatively low reliability to serve its purpose. For instance, continuing the previous numerical example, if a standby system with a failure probability as poor as 10^{-2} were placed in parallel with the main system, then the total failure probability would be,

$$\text{P (main system)} \times \text{P (standby system)}$$
$$= 1.01 \times 10^{-7} \times 1.0 \times 10^{-2}$$
$$= 1.01 \times 10^{-9}$$

which is a satisfactorily high reliability.

Unwanted Operation of Systems

So far in this Chapter, failures have been taken as a failure to function when needed. However, the opposite problem sometimes arises. A system can operate, and cause danger by so doing, when it is not needed to operate. An example is the stick-pusher, where an unwanted operation near the ground could be hazardous. Likewise the unwanted operation of warning systems, when the aircraft is in fact behaving normally, is objectionable and can be dangerous.

The usual design practice to ensure a high degree of certainty that the system will operate when wanted, is by the use of multiplex channels in parallel. However, when it is important to avoid unwanted operation, items may be put in series. For instance, in the stick-pusher example, two sensors might be used, both indicating the need for the stick to push. If the probability of either sensor giving a false signal is P, the probability of both combining to give a false signal would be P^2.

Again the arithmetic follows the same lines, but practical judgment is necessary to balance correctly the risks of failure to operate when wanted against the risk of operating when not wanted.

Where 'P' is Not a Small Quantity

Before leaving these pages dealing with combinations of probability, it is as well to recall the simplifying assumption that where P is sufficiently small $(1-P)$ is virtually equal to 1. While this is usually the case in aircraft system problems, it is not invariably so. To deal with the general case, where P can be any number between 0 and 1, let us write $Q = 1-P$. Then for two components the combined probabilities are:–

$$
\begin{aligned}
&\text{A and B fail} && P^2 \\
&\text{A fails; B operates} && PQ \\
&\text{B fails; A operates} && PQ \\
&\text{A and B operate} && Q^2
\end{aligned}
\quad \left.\begin{matrix} \\ \\ \end{matrix}\right\} 2PQ
$$

Extending this to larger numbers leads to the following table:–

No. of Channels	No. of Failures			
	1	2	3	4
1	P	—	—	—
2	2PQ	P^2	—	—
3	$3PQ^2$	$3P^2Q$	P^3	—
4	$4PQ^3$	$6P^2Q^2$	$4P^3$	P^4

For the mathematically inclined reader, the terms in the above Table are the terms of the binomial expansion of $(P + Q)^n$.

THE EFFECTS OF TIME

Probability – Per Flight and Per Hour

So far probability has been written as 'P' meaning the probability of the occurrence of an event.

Now some risks are dependent on the number of hours of exposure to risk. One practical result of this should be noted at once. If the failure probability of a channel is p per hour, then the probability of failure in a flight duration T hours will simply be pT. (Strictly this is an approximation, as is discussed in later pages). It follows that the probability of a double failure of a duplicated system is p^2T^2; and of a triple failure of a triplicated system is p^3T^3.

Note that the duration of flight has a significant effect on the risk per flight. For example, if $p = 10^{-3}$ per hour, the probability of failure **per flight** is,

	T = 1	T = 5
Single	10^{-3}	5×10^{-3}
Double	10^{-6}	25×10^{-6}
Treble	10^{-9}	125×10^{-9}

The longer the duration the lower the reliability, but this is far more pronounced when considering multiplex systems.

A small warning should be mentioned here. While many risks are dependent on the number of hours of exposure, others depend on the number of times an item operates. For instance, the latter would include the landing gear mechanism which normally is used twice per flight. In these cases, if the probability 'per use' is p, then the probability in a flight is np, where n is the number of uses of the item per flight. Thus when estimating total probabilities, including probabilities of combined failures, it is important to remember whether the individual probabilities are of the "per hour" kind or the 'per use' kind.

Relationship Between Probability and Mean Time Between Failure

Information on the reliability of components is often available in the form of the Mean Time Between Failure (MTBF). It is needed to find the probability per hour which corresponds with the MTBF.

If the failure rates are plotted against time since the component was last overhauled, various pictures emerge. Usually, over the period between overhauls, the rate of failure is found to be reasonably constant. In other words, the failures crop up in a random fashion, unrelated to newness or oldness. In such cases, it is clear that the probability per hour is simply the reciprocal of the MTBF. Thus,

$$p = \frac{1}{T_m} \text{ where } T_m \text{ is the MTBF.}$$

However, where a component deteriorates with time, due to fatigue or wear, the probability of failure plainly increases with age. In most cases of aircraft system components, the practice would be to remove the component for overhaul or replacement before the onset of this increasing risk. Hence the simple $p = 1/T_m$ rule would still apply.

The converse situation can also arise in which an item is more prone to fail shortly after overhaul than later on. This is not often the case, but should it be so the failure probability can be conservatively chosen so as to allow for the increased initial risk.

Thus, for most practical cases we can take p as being equal to $1/T_m$. Since to a close approximation the probability of failure in t hours equals pt, the risk of failure in t hours can be written as t/T_m. The closeness of this approximation is discussed below.

If the failure records of a component show that there are different modes of failure occurring with different effects, then it may be necessary to calculate the MTBF for each separate mode. For instance one mode might cause the function to cease, and another mode might cause unwanted action, with different effects.

The Exponential Curve

A somewhat complex piece of arithmetic shows that if the probability of failure per hour is p then the risk of failure before t hours is strictly,

$$1 - e^{-pt} \quad \text{or} \quad 1 - e^{-t/T_m}$$

where e is the exponential number 2·718.

The reason for this rather curious looking result can more readily be seen by considering not just one item but several. Suppose that 1,000 items, each with a failure probability per hour of 1/1,000, start to run simultaneously. By the time that each has run, or tried to run, for 100 hours, a total time of approximately 100,000 hours will be reached. In this period, 100 items will fail. The surviving 900 items then run on for another 100 hours and 90 will fail. The surviving 810 items will lose 81 more in the next 100 hour period, and so on. In short, as the number of surviving items diminishes, so also will the number of failures. If the number of failures is plotted against total hours a curve of the shape shown in Fig. 5-1 results. This is in the form $1 - e^{-pt}$. It will be seen that out of the initial large number of items 63% fail before the MTBF is reached, and 37% survive.

On exactly the same reasoning, it can be shown that an individual item has a probability of failing before time t equal to $1 - e^{-pt}$. This "exponential failure curve" appears frequently in safety analysis.

To return to the typical aircraft system problem, if we are considering the active failure of a component in a flight, then the order of numbers is a flight time t of 10 hours (or less) and a probability p of 10^{-3} per hour. Thus pt is unlikely to exceed 0·01. For such a small value of pt, to a very close approximation $1 - e^{-pt} = pt$. This is indicated in Fig. 5-1. Thus for many calculations the failure risk in a flight of t hours can be taken simply as pt.

The value of $1 - e^{-pt}$ can be written as an expansion, as follows,

$$1 - e^{-pt} = pt - \frac{p^2t^2}{2} + \frac{p^3t^3}{6} - \frac{p^4t^4}{24} \dots \text{etc.}$$

Thus, if there is any doubt whether pt is sufficiently small to justify the above approximation, this series provides an easy way of checking.

In particular, the approximation may not be accurate enough in cases of 'dormant faults' as discussed later.

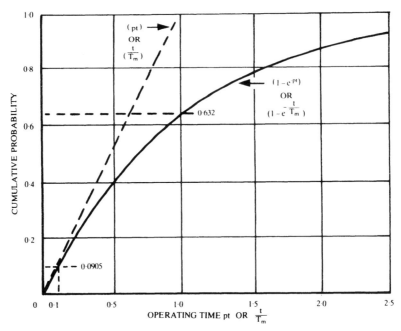

Fig. 5-1   THE EXPONENTIAL CURVE

In the above simplified explanation of the Exponential Curve, the term 'probability per hour' was used. Objection can be made to this on the grounds that probability is non-dimensional. A more precise statement of the underlying assumption is that the probability of failure in a short interval of time, dt, equals p dt, where p is constant and not dependent on the length of time t that the item has run. The words 'probability per hour' can therefore be regarded as a shorthand way of writing 'probability of failure in a period of one hour', which for all practical purposes is p.

In statistical literature reference is sometimes made to "probability density". This need not concern us but it is mentioned here to avoid confusion. Probability density is in effect the frequency of occurrence of failures at a particular time, t. As time proceeds and failures occur the number of surviving items diminishes, as was illustrated previously by a numerical example. With fewer survivors there are fewer failures. For a constant probability of failure p, the probability density falls away in accordance with the diminishing curve, $p\,e^{-pt}$.

This distinction between probability per hour and probability density must be kept clearly in mind when using service records to establish the pattern of failures versus time. The simplest approach is to count failures occurring between time intervals (e.g. 0 to 500 hours, 500 to 1,000 hours) and to divide each by the corresponding total number of item-hours in each interval. If this gives a more or less constant figure, the probability per hour can be taken as constant, and any confusion with the varying probability density avoided.

Dormant Faults

In some systems there can be a fault in one channel which leaves the system operating, and the presence of the fault is undetectable by the pilot. Such faults, called 'dormant faults' are only revealed when other channels fail. The same problem may apply to standby or warning systems, which are only required to operate infrequently. Even though between operations such systems are doing nothing, an undetected dormant fault could still occur and make its presence felt when the system should have operated.

The risk that dormant faults are present can be mitigated by making specific checks, e.g. before each flight or at check periods. If the probability of a dormant fault is p per hour, and a check is made every T_c hours, then the probability of a dormant fault being present is $p \times T_c$. If T_c is a long period comparable with the MTBF, then the more precise formula $1 - e^{pT}$ must be applied.

There is a further, significant, point which arises from this 'time effect'. Previously, it was shown that multiplexing the number of channels is an extremely powerful way of reducing the chance of total system failure. The example quoted was for a single channel failure probability of 10^{-3}, and this gave a double failure probability of 10^{-6}, a very large gain indeed. However, when the single channel failure probability is itself high, then the gain to be had from multiplexing is very much less.

The reason is simple. If the single failure risk is pt, the double failure risk is $(pt)^2$, and the treble failure risk is $(pt)^3$, and so on. If pt is small, then $(pt)^2$ is tiny. But if pt is large, and nearly equal to 1, then $(pt)^2$ is also nearly equal to 1.

Fig. 5-2 illustrates this point in general terms, showing a plot of failure probability against t/T_m. For the single channel case, the probability is,

$$1 - e^{-t}/T_m$$

and for the double channel, double failure case is,

$$\left(1 - e^{-t}/T_m \right)^2$$

The practical implications can be seen from an example. Suppose that the aim is to limit the risk of being inoperative due to dormant failures to 0·5, and to make checks for the existence of dormant failures at intervals of t = 1,000 hours. Then reading from the curves of Fig. 5-2,

	t/T_m	Required MTBF
Single	0·7	1430
Double	1·2	830
Treble	1·6	620

Thus, though multiplexing permits a reduction in the required MTBF, the gain is a small one. The lesson is that when the issue involved is dormant failures and long intervals between checks, multiplexing is of no great help. The better solution is the provision of basically higher channel reliability and/or reduced periods between checks.

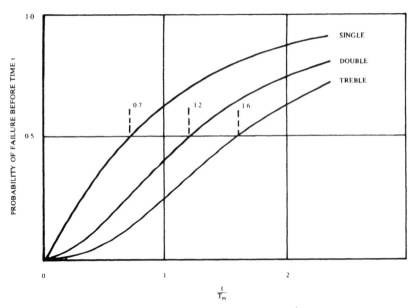

Fig. 5-2 FAILURE PROBABILITIES WHERE $\dfrac{t}{T_m}$ IS "LARGE"

Where the Sequence of Failures is Important

In the previous examples, it has been assumed that the sequence of failures in the flight makes no difference to the end result. This is often the case but not always.

For instance, the activating of a standby might depend on a warning that the main system had failed. If the warning was already inoperative by virtue of a dormant fault, the operability of the standby would be of no consequence, as the pilot would be unaware of the need to activate it. So in some instances the order of failures is significant.

As has already been stated, with two items, A and B, if the sequence of failures is immaterial, the probability of double failure is,

$$p_A p_B \, T^2$$

Since approximately half the failures involve A failing before B, and vice versa, the probability in a specified order is,

$$p_A p_B \, \frac{T^2}{2}$$

This can be demonstrated more exactly as follows. Consider the sequence of failure A occurring before failure B. The probability of A failing before a time t is $1 - e^{-p_A t}$; of B not failing before time t is $e^{-p_B t}$; and of B failing in the interval between t and (t + dt) is $p_B \, dt$. Over the whole flight of duration T, the probability that A will fail before B is therefore:–

$$\int_0^T (1 - e^{-p_A t}) \, e^{-p_B t} \, p_B \, dt$$

52

This works out, to a first approximation, to be:–

$$\frac{p_A p_B T^2}{2} \left(1 - \frac{(p_A + 2p_B)}{3} T\right)$$

The sequence of B occurring before A is similarly:–

$$\frac{p_A p_B T^2}{2} \left(1 - \frac{(p_B + 2p_A)}{3} T\right)$$

Thus, when the terms $p_A T$ and $p_B T$ are small, the probabilities of A before B, and B before A, are to a close approximation the same and equal to:–

$$p_A p_B \frac{T^2}{2}$$

The same argument can be applied to say 3 items. The order of failures can be ABC, ACB, BAC, BCA, CAB, or CBA. Thus one-sixth of the failures are in a specified order with a probability of,

$$p_A p_B p_C \frac{T^3}{6}$$

Where the Flight Procedure is Important

In some cases of serious first failure, the flight procedure may be varied to reduce the risk of a second failure. For example, after a first engine failure on a twin-engined aircraft, a decision may be made to return to base, or to divert to an alternate.

In such instances, the probability of multiple failures will depend on the strategy of the flight. Take as an example two items, each with a failure probability p per hour, on a flight with an intended duration of T hours.

(a) The simplest strategy is to **continue** the flight after the first failure. As stated previously, the probability of double failure is,

$$p^2 T^2.$$

(b) A possible strategy would be to **return to base** after the first failure. The probability of one item failing in a short time interval is p dt. If this occurs at time t, the return flight can be assumed to take t hours, so the probability of a second failure is pt. The probability of a double failure over the whole flight is thus,

$$\int_0^T p^2 t \, dt = p^2 \left(\frac{t^2}{2}\right)_0^T = p^2 \frac{T^2}{2}$$

As the sequence can be either AB or BA, the total probability is:–

$$p^2 T^2$$

which is the same as the 'continue to destination' case, (a).

(c) A more beneficial strategy would be to return to base if the first failure occurred before time T/2 and to continue to the destination if the first failure occurred after time $T/2$. We can regard the flight as being in two halves, each of duration $T/2$. From (a) and (b) above, the probability of double failure in each half is,

$$p^2 \left(\frac{T}{2} \right)^2$$

The total probability is, therefore,

$$p^2 \frac{T^2}{2}$$

This strategy therefore halves the total risk as compared with either (a) or (b).

(d) On long flights the risk can be reduced by diverting to an alternate after the first failure. Suppose that there was always an alternate within H hours flight time of the track. The probability of a first failure in an interval of time dt is as before p dt, and the probability of a second failure in the remaining flight time H is pH. The double failure probability is thus,

$$\int_o^T p^2 H \, dt = p^2 H \left(t \right)_o^T = p^2 H T$$

Again, as the sequence can be either AB or BA the total probability is,
$$2 \, p^2 \, H \, T$$

As an example of the use to which the foregoing can be put, suppose that an aircraft had a duplicated system, the single failure of which had a probability of 10^{-4} per hour. Suppose further that the consequence of double failure had a sufficiently hazardous effect to limit the risk to 10^{-8} per hour.

Then, if the flight procedure were to fly on to the destination after a first failure, the double failure probability is p^2T^2 which leads to,

T	1	2	4	8	
p^2T^2	1	4	16	64	$\times 10^{-8}$
Permissible	1	2	4	8	$\times 10^{-8}$

With this procedure only short flights of up to 1 hour would be permissible.

If the procedure were to return to base following a first failure up to the mid-point of the flight, and to proceed to the destination if the failure occurred after the mid-point, the probability is $p^2\dfrac{T^2}{2}$ leading to,

T	1	2	4	8	
$\dfrac{p^2T^2}{2}$	½	2	8	32	$\times 10^{-8}$
Permissible	1	2	4	8	$\times 10^{-8}$

This strategy permits flights up to 2 hours.

For longer flights, it would be necessary to plan to use alternate landing points. Suppose that it were possible to divert to an alternate not more than ½ hour from the flight track. The procedure would then be, after the first failure, to return to base in the first ½ hour of flight, to continue to destination in the last ½ hour, and in the central segment (duration T − 1 hours) to divert to the nearest alternate. The double failure probability is then,

$$p^2(½)^2 + 2p^2 \ ½ \ (T − 1) + p^2(½)^2 = p^2 \ (T − ½)$$

This leads to,

T	2	4	8	
$p^2(T − ½)$	1½	3½	7½	× 10^{-8}
Permissible	2	4	8	× 10^{-8}

With this last strategy, no matter how long the flight, the double failure risk is within the permissible limit.

DETERMINING RATES OF FAILURE FROM SMALL SAMPLES
The 'Expected' Number

When there are large amounts of data, the total number of failures divided by the total running time gives a failure rate in which we can feel confident.

However, it may well be that relatively small amounts of information are all that is available, and the question then arises of making the best use of what we have. For instance, at the design stage it may be necessary to establish the MTBF from comparatively short test runs.

To illustrate the point, suppose that the MTBF of an item is T_m, then if a run of $5 \times T_m$ is made one could 'expect' that just 5 failures would occur. But it would be no surprise if in fact, because of chance variation, fewer than 5 or more than 5 failures actually turned up. Thus, when T_m is not known from long experience, short runs are poor guides to the likely value of T_m. However, there is a handy statistical tool, called the Poisson Distribution which helps with this problem.

The Poisson Distribution

The Poisson Distribution shows that if the 'expected' number of failures is n, then the probability P that F actual failures will occur is as follows,

F	0	1	2	3	F
P	e^{-n}	$e^{-n}n$	$e^{-n}\dfrac{n^2}{2!}$	$e^{-n}\dfrac{n^3}{3!}$	$e^{-n}\dfrac{n^F}{F!}$.

In this numbers such as 2! and 3! are factorial numbers: $2! = 2 \times 1$; $3! = 3 \times 2 \times 1$; etc.

It is easier to grasp the practical significance of this Poisson Distribution by applying it to typical numbers. Fig. 5-3 shows the probabilities of F failures for various values of the expected number, n.

First, it is apparent that the most likely result is that the number of failures actually occurring will equal n or n − 1. These two results are the most probable. Second, though these two results are the most probable, it is not very improbable that fewer or greater numbers will appear.

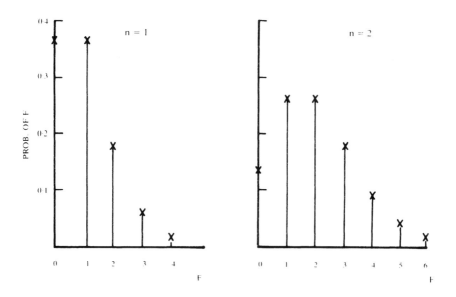

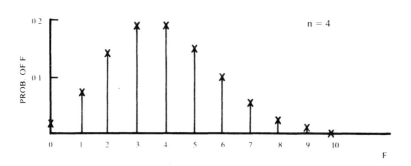

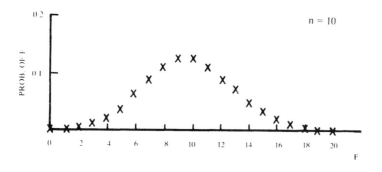

Fig. 5-3   EXAMPLES OF POISSON DISTRIBUTION

56

It is quite clear that we cannot learn much from small samples. For instance, in a run of $2 \times T_m$ we would expect 2 failures. If only one failure actually occurred, could we assume that the item was behaving twice as well as expectation? This would be a rash conclusion. When $n = 2$, it is equally likely that 1 or 2 failures will actually occur. Further it is not very unlikely that no failures will occur.

In the other direction, when $n = 2$, a much larger number, say 5 failures might occur. The sum of the probabilities of $F = 5$ or more failures is around 0.1. So the odds are about 10 to 1 against such a result occurring from pure chance.

However, to have any real confidence in the results, more evidence is necessary. For example, suppose that an item is believed to have an MTBF of 1,000 hours, and that in service 15 failures occur in a period of 10,000 hours. This period equals $10 \times T_m$, and hence the expected number of failures is $n = 10$. Summing the probabilities of 15 or more failures, when $n = 10$, shows that the likelihood of this result is about 0.07. Thus we could have a confidence of over 90% that the appearance of 15 failures when 10 were expected was not a pure chance event. We would have reason to believe the service record was worse than expectation.

It is clear from the above that from samples of service experience we cannot determine with absolute **certainty** that a component is better or worse than expectation, but we can attach a numerical **confidence** to our belief.

Initial Choice of MTBF

Suppose that for a new item all we know is that F failures have occurred in H hours, indicating experimentally that MTBF is H/F. What we need to decide is the higher number of failures N which might occur in H hours, in the long term. We can then decide with a certain level of confidence that the MTBF will not be less than H/N.

Text books on statistics give tables or curves showing the Poisson relationship between F, N, and confidence level. Figs. 5-4(a) and (b) have been derived from these generalised values. Note that though N is plotted against F as a continuous curve, in fact F must be a whole number, 1, or 2, or 3, etc.

To illustrate the use of these figures, suppose we wanted to be fairly confident in the interpretation of test results, we could use the curve for 80% confidence, this meaning that there would be a 4 to 1 chance that we were on the safe side. If, in a test run of 8,000 hours, $F = 5$ failures actually occurred, the 80% confidence curve shows a value of $N = 8$. Thus we would convert the experimental MTBF of 1,600 hours to the more conservative figure of 1,000 hours.

For a run of say 20,000 hours in which 20 failures actually occurred, the 80% confidence curve shows a value of $N = 25$. Thus we would convert the experimental MTBF of 1,000 hours to a conservative one of 800 hours.

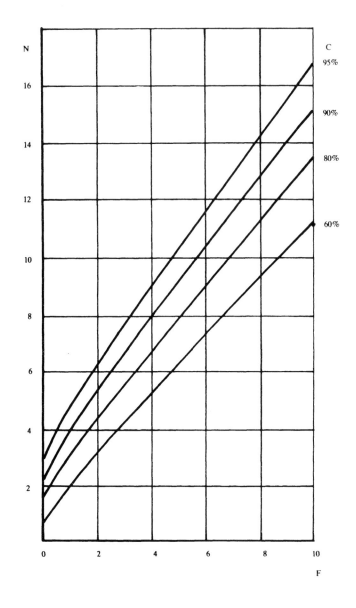

**Fig. 5-4(a)   RELATIONSHIP BETWEEN N, F AND
CONFIDENCE LEVEL**

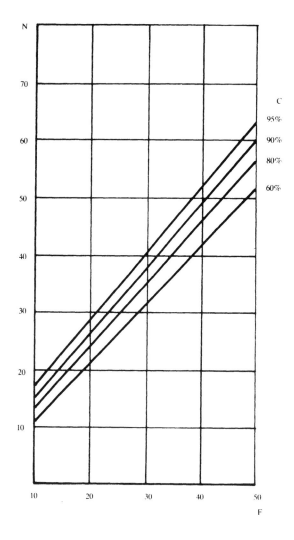

Fig. 5.4(b)   RELATIONSHIP BETWEEN N, F AND
CONFIDENCE LEVEL

59

It is to be noted that the higher the degree of confidence wanted, the larger is the factor to be applied. Also it is evident that the fewer the actual failures, the larger the multiplying factor.

PERFORMANCE OF SYSTEMS

Variability of Performance

In addition to outright failure to function, there is often the need to examine the effects on safety of components operating but at reduced levels of 'performance'. The word performance is used here in a wide sense; in engines, it could mean thrust; in electrical components, an appropriate measure of output; in instruments, their accuracy.

It is plainly unlikely that all components of a given type will always produce exactly the same 'performance'. It is more likely, and tests show this to be so, that performance is distributed about some mean value.

A safety assessment may, therefore, be needed to examine the degree of degradation of performance which might be encountered. For this purpose we need to know something about the distribution; i.e. the shape and spread of the scatter about the mean.

To begin with a fairly typical example, Fig. 5-5, shows in histogram form 100 measurements of engine thrust. To simplify the presentation, the mean value of thrust determined from the test results is expressed as $\overline{T} = 1{\cdot}0$. The figure shows the numbers of results which fall in brackets of thrust, e.g. between 0·985 and 0·995. The shape of the diagram shows a fair degree of symmetry about the mean, and also shows that as one gets further from the mean, above or below, the proportion of cases reduces.

This diagram also suggests the possibility that, if instead of 100 test results, 1,000 were obtained, even more extreme high or low values would appear. Unless there is some external intervention to limit the scatter, it seems that on less and less frequent occasions more and more extreme values will occur.

Where there is no clear absolute maximum or minimum, we must find the value of the measured characteristic, in this case thrust, above which or below which the performance might be on, say, 1 in 100 or 1 in 1,000 occasions. The raw data itself provides some clue. For example in this case 2 tests out of 100 showed a thrust of 0·955, a degradation of about 5% below the mean. To obtain a better interpretation of the data, we could draw a curve of the kind shown in Fig. 5-5 to smooth out the irregularities of the test results.

It is a curious fact that many aspects of performance, and many physical characteristics such as strength of materials, are distributed in accordance with a smooth curve, known as the Normal (or Gaussian) Distribution. The curve shown in Fig. 5-5 is in fact a Normal Distribution curve.

It is not easy to account for this fact that naturally occurring variation often follows this particular shape. For present purposes, it is accepted that for many of problems encountered in practice, the Normal curve is a fair approximation to reality. The question is then how to put this fact to use.

60

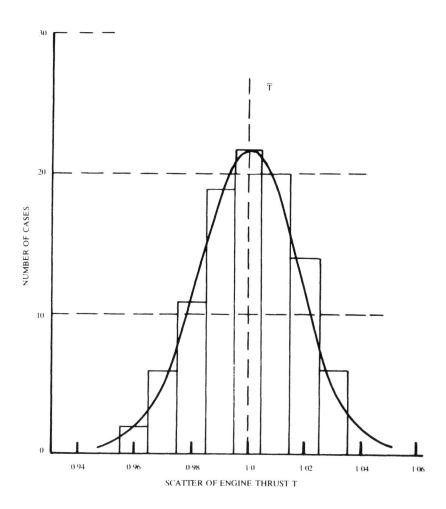

NUMBER OF CASES

SCATTER OF ENGINE THRUST T

Fig. 5-5   SCATTER OF ENGINE THRUST (T)

The Normal Curve

It would, theoretically, be possible, starting with a set of test results such as shown in Fig.5-5, literally to draw the Normal curve which was the best fit of the test results. This, however, would be laborious and unnecessary as there is a quicker approach.

The Normal curve is completely defined by two numerical values:–
a. the ordinary mean value of the characteristic being measured, and
b. what is known as the standard deviation (s.d.) which is in effect a measure of the width of the scatter.

Thus, if we calculate, for our test results, their mean and standard deviation, we arrive at the Normal curve which best fits the test data. Note that there are two underlying assumptions; that the sample of results is large enough to be typical of the whole 'population', and that the population is in fact a Normal one.

The calculation of the mean is straightforward. The calculation of the standard deviation involves finding the 'root mean square' of the individual results relative to the mean. The way in which this is done is best seen from a worked example. The previous thrust test results are written out in Table 5-1. The first column shows the results grouped into various levels. The second column shows the number of cases falling within each bracket of thrust. The third column shows the departure of the individual measured values from the overall mean thrust, and the fourth column gives the square of these figures. The last column, when added, gives the sum of the squares of the differences of the individual results from the mean. The standard deviation is then the square root of the mean of the sum, viz.

$$\text{standard deviation} = \sqrt{\frac{\Sigma \, n(T - \overline{T})^2}{\Sigma \, n}}$$

TABLE 5-1
VARIATION OF ENGINE THRUST

Measured thrust (T)	No. of cases (n)	$T - \overline{T}$	$(T - \overline{T})^2$	$n(T - \overline{T})^2$
0·955 — 0·965	2	− 0·04	·0016	·0032
0·965 — 0·975	6	− 0·03	·0009	·0054
0·975 — 0·985	11	− 0·02	·0004	·0044
0·985 — 0·995	19	− 0·01	·0001	·0019
0·995 — 1·005	22	0	0	0
1·005 — 1·015	20	+ 0·01	·0001	·0020
1·015 — 1·025	14	+ 0·02	·0004	·0056
1·025 — 1·035	6	+ 0·03	·0009	·0054
	$\Sigma n = 100$			$\Sigma n(T - \overline{T})^2$ $= 0 \cdot 0279$

$$\text{s.d.} = \sqrt{\frac{\Sigma \, n(T - \overline{T})^2}{\Sigma \, n}} = 0 \cdot 0167$$

In most text books on statistics, the properties of the Normal curve are fully tabulated. Fig. 5-6 illustrates a few relationships, e.g.,

10% of all cases lie below a level 1·3 × s.d. from the mean
1%   — ditto —   2·3 × s.d. — ditto —
0·1%   — ditto —   3·1 × s.d. — ditto —
0·01%   — ditto —   3·7 × s.d. — ditto —

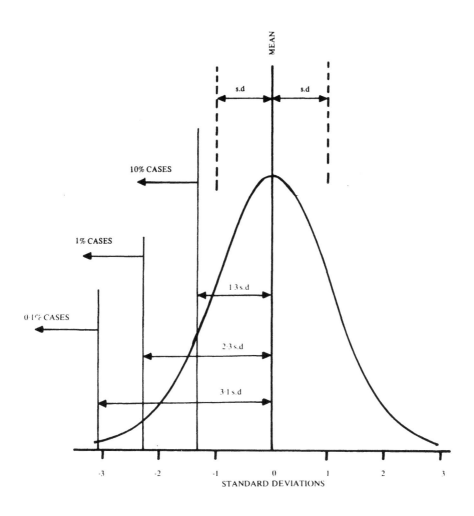

Fig. 5-6   **PROPERTIES OF THE NORMAL CURVE**

We can now apply the above to the engine thrust example to discover the reduction of performance corresponding to various frequencies of occurrence, as shown below.

Proportion	Corresponding Probability	Reduction Below Mean Thrust	Reduced Level of Thrust
1%	10^{-2}	2.3×0.0167	96.2%
0.1%	10^{-3}	3.1×0.0167	94.7%
0.01%	10^{-4}	3.7×0.0167	93.7%

These results give a better approximation to the degree of degradation of performance than can be obtained from simple visual inspection of the original set of test results.

A Word of Warning

When, as in the last example, we use the Normal curve as a means of finding the level a characteristic falls to on rare occasions, it should always be remembered that we are in fact extrapolating outside the range of the measured values. All that we can be wholly certain about is what has been measured. Any form of extrapolation assumes that the same pattern of distribution applies beyond the range of the measured values. Gross extrapolation can be misleading.

The 'tails' of the Normal Distribution never quite reach zero. For example, the Normal curve suggests that in one case in 10 million, the particular characteristic would fall to 5.2 standard deviations below the mean. If we were to be tempted to conclude from this that on 1 in 10^7 occasions thrust would fall to 91.3% of the mean, we would probably be abusing the statistical method. In practice, if thrust were to fall to exceptionally low values, it would become apparent, and steps would be taken to correct it. Thus, because of such intervention, the Normal curve no longer applies. In effect, the tails are trimmed off.

Therefore, when applying the Normal curve procedure in the way described above, it can safely be used to extrapolate in a moderate way; but gross extrapolation should be regarded with reserve.

This point becomes very apparent if characteristics such as approach speed are examined. Measured data of approach speed may well suggest that over a fair range of speeds the variation is Normal. However, it is quite clear that it would be false to deduce from this that on rare occasions the approach speed was less than stalling speed.

What To Do About Small Samples of Measurements

When the number of tests made is fairly large, it is reasonable to suppose that they are a representative of the whole 'population'. Thus the calculated mean and s.d. will be good approximations to the true mean and s.d. of the characteristic measured. Given say 30 results we can be reasonably confident, and given 100 results very confident that they are representative.

But we may be obliged to use a much smaller test sample comprising say 5 or 10 results. With such a small number there is a large element of chance whether they are typical or not.

There is first the question whether the characteristic being tested conforms to the Normal Distribution. The previous example of engine thrust showed how a sample of 100 results gave a fairly clear indication that it fitted the Normal Distribution. If tests are made on a new design of component, but components of generally similar design are known from past experience to conform to the Normal Distribution, it is reasonable to assume that the characteristic of the new design will be Normal, even though the number of tests on the new design is not itself sufficient to prove the fact.

Where we believe, either from past experience or from a large enough number of specific tests, that we are dealing with a characteristic which follows the Normal Distribution, there still remains the question of the accuracy of the mean and standard deviation which results from the test sample. It is evident that the larger the number of test results, the greater the likelihood that the answers will be accurate representations of the 'true' mean and s.d.

To err on the safe side, it is generally preferable to assume that true mean and true s.d. will be a little different from the calculated results. There are statistical methods for making the appropriate adjustment.

These methods start with the concept of what is called the "standard error". Numerically, this is,

$$\text{for the mean} \quad \frac{\text{s.d.}}{\sqrt{n}}$$

$$\text{for the s.d.} \quad \frac{\text{s.d.}}{\sqrt{2n}}$$

where n is the number of results in the sample.

If we were concerned with the lowest likely performance we would adjust the calculated mean downwards and would increase the calculated s.d. By applying an adjustment of **twice** the above standard errors, we would be 95% certain that the adjusted values would not be optimistic.

Examples of the adjustments for different sample sizes are given below.

	n	25	100	400
To the mean	$2 \times \dfrac{\text{s.d.}}{\sqrt{n}}$	0·4	0·2	0·1 × s.d.
To the s.d.	$2 \times \dfrac{\text{s.d.}}{\sqrt{2n}}$	0·28	0·14	0·07 × s.d.

The effect of applying these adjustments can be seen by way of example. Suppose that the test results indicated a mean of 100 units, and a s.d. of 4 units. Then the reduced mean and the increased s.d. resulting from applying these adjustments would be as follows,

n	25	100	400
Mean	98·4	99·2	99·6
s.d.	5·1	4·6	4·3

It will be seen that the adjusted mean is still very close to the original calculated value of 100. On the other hand the adjusted s.d. is appreciably increased relative to the original calculated value of 4.

The value of having a large sample of test results is clear, and this is particularly so if the characteristic concerned affects safety in a critical way.

Strictly speaking, the above method of adjusting test results is applicable only to reasonably large samples. For small samples there are more elaborate statistical methods, (the Student t test and the chi-square test). However these are rather beyond the scope of this Chapter which is intended to cover the simpler aspects.

Combination of Variation

A system may comprise a number of components, each with its particular degree of variation and each contributing to the variation of the total output of the system.

This combination of variability is dealt with as follows.

If there are components with standard deviations $s.d._1$, $s.d._2$, $s.d._3$, and so on, the combined standard deviation is,

$$\sqrt{s.d._1{}^2 + s.d._2{}^2 + s.d._3{}^2}.$$

If for instance there are three components with s.d's of 2, 3, and 4, respectively, the resulting combined s.d. is,

$$\sqrt{4 + 9 + 16} = 5·4.$$

Further Remarks on Distribution

In earlier paragraphs, dealing with probability of failures, it was noted that many failures of systems are often of a kind which are equally likely to occur irrespective of the newness or oldness of the component. This permits much simplification in the analysis. In subsequent paragraphs dealing with 'performance', it was noted that many qualities of systems follow the Normal Distribution, again easing the arithmetic.

It must be emphasised that these assumptions should not be taken as foregone conclusions. Faced with anything new, it is as well to examine whether these assumptions are justified. It may be that some other distribution suits the case better.

One particular example of this may be mentioned here. Structural fatigue failures are obviously connected with time since new. The probability of failure quite clearly increases as the part accumulates flying time. If fatigue tests are made, they show that there is considerable variation of the measured life of a component about its mean life. The problem of establishing a safe life is therefore rather like establishing the variation of 'performance' described previously. One is concerned with finding a life which will not be undercut more often than once in X occasions.

It has been found empirically that fatigue lives are scattered about the mean in what is called the 'log normal distribution'. Again this is not a fundamental law, but simply a useful practical rule of thumb. The 'log normal distribution' is simply this. If the logarithms of fatigue lives measured for a particular item are plotted, then their distribution about the mean follows a Normal Distribution. The results can, therefore, be interpreted using the known properties of the normal curve, as previously described.

Useful References
1 "Facts from Figures", M. J. Moroney, Penguin Books.
2 "Reliability Theory in Practice", Igor Bazovsky, Prentice Hall Space Technology Series.

APPENDIX 5-1
ASSESSMENT OF A SIMPLE SYSTEM

Introduction

Consider a simple system which comprises a Main System, M; a Warning, W, when the main system fails; and a Standby system, S. The Safety Assessment could best be performed as follows:

Step 1

The first step is to write a description of the way the system behaves.

The main system is in continuous operation in flight.

If it is inoperative at the start of flight, this is evident to the pilot, and the instructions are to cancel the flight.

If it fails in flight, this is not sufficiently evident to the pilot, unless the warning system operates.

On seeing the warning, the instructions are to check the functioning of the main system and if it has failed, to switch on the standby system.

The standby and warning can only be checked on the ground, such checks being required to be made at specified intervals. The instructions are to rectify faults before further flight.

Step 2

The second step is to state the hazardous condition for which a probability has to be established. In this case, it is the total loss of system function in flight.

Step 3

The third step is to assemble the numerical values needed for the assessment:–

Flight duration	T	$= 2$ hrs.
Check period of standby	T_s	$= 100$ hrs.
Check period of warning	T_w	$= 50$ hrs.
Probability of failure of main system when active	p_m	$= 10^{-4}$ per hr.
Probability of failure of standby system when active	p_{sa}	$= 10^{-3}$ per hr.
Probability of dormant failure of standby when inactive	p_{si}	$= 10^{-5}$ per hr.
Probability of dormant failure of warning	p_w	$= 10^{-4}$ per hr.

In this example, the probability of human errors, (failure to notice or act on the warning, failure to make the ground checks) will not be considered. We are, therefore, concerned with combinations of dormant failures in the standby and warning at the start of the flight, and with failures of operating systems during their use in flight.

It is assumed that the warning will be recognised by the pilot as soon as it operates so that it is not necessary to consider failures of the warning while it is active.

68

Step 4

The fourth step is to write down all the possible circumstances (states of the system elements) for the hazardous condition to be present.

In this case, the main system must be operating before take-off, and it is assumed that the pilot will only switch over from it if it ceases to operate (i.e. false warnings of failure can be neglected). Therefore, the first event that must occur is a failure of the main system. But there are two ways in which this can lead to loss of system function:–

A. If the pilot fails to switch over to the standby.
B. If the pilot does switch over to the standby, but there is an existing or subsequent loss of the standby.

Step 5

The fifth step is to define the precise failure sequences which can lead to the hazard.

A. Since it is assumed that the pilot will always see the warning, the only reason for not switching over when the system fails is if the warning is inoperative.

   The probability per flight of the main system failing is $p_m T$.

   The probability that the warning is failed will depend on the time that has elapsed since it was checked, i.e. the probability that it has failed is sensibly zero in the instant after the check, and is $p_w T_w$ immediately before the next check. So that the average risk that the warning is inoperative is $\dfrac{p_w T_w}{2}$.

NOTE: A check should be made that terms such as $p_w T_w$ are sufficiently small that this approximation to $1 - e^{-pt}$ is valid.

The probability of the fault sequence "warning failure" followed by "main system failure" is, therefore:–

$$p_m T \times \frac{p_w T_w}{2} = 10^{-4} \times 2 \times \frac{10^{-4} \times 50}{2} = 5 \times 10^{-7} \text{ per flight.}$$

B. In this case where the hazard arises from loss of main system and loss of standby, there are two fault sequences to consider:–

   Case B_1. Where the standby system is already inoperative when the pilot switches over to it. This case is exactly equivalent to the dormant failure of the warning: i.e. the probability of the fault sequence "dormant failure of standby" followed by "failure of main system" is:–

$$p_m T \times \frac{p_{si} T_s}{2} = \frac{10^{-4} \times 2 \times 10^{-5} \times 100}{2} = 10^{-7} \text{ per flight.}$$

69

Case B_2.   Where the standby system fails after the pilot switches over to it. Following the rules given earlier in this Chapter for failures of two items happening in a particular sequence, the probability of "main system failure" followed by "standby system failure when active" is,

$$p_m p_{sa} \frac{T^2}{2} = 10^{-4} \times 10^{-3} \times \frac{4}{2} = 2 \times 10^{-7}.$$

Step 6

The sixth step is to sum the risks, i.e., risk per flight is,

$$A + B_1 + B_2 = (5 \times 10^{-7}) + (10^{-7}) + (2 \times 10^{-7}) = 8 \times 10^{-7}.$$

And, since average flight time is 2 hours, the risk per hour is 4×10^{-7}.

Step 7

The seventh step is to consider the acceptability of the assessed probability. If it is too high in relation to the consequences of the failure condition then it is necessary to consider what action should be taken.

For example, if loss of system function were considered so hazardous that the probability was unacceptable, then action could be taken, as follows:–

Clearly the probability of the main system failure contributes directly to the average risk through all three failure sequences. Thus any improvement in its reliability would be reflected proportionately in the total.

The greatest contribution to the risk arises from case "A", and this contribution could be reduced if there were a pre-flight test of the warning system.

It would become $p_m p_w \dfrac{T^2}{2} = 10^{-4} \times 10^{-4} \times \dfrac{4}{2} = 2 \times 10^{-8}$ per flight $= 10^{-8}$ per hour.

Similarly the contribution of the dormant failure of the standby system through case B_1 could be reduced by a pre-flight check.

It would become $p_m p_{si} \dfrac{T^2}{2} = 10^{-4} \times 10^{-5} \times \dfrac{4}{2} = 2 \times 10^{-9}$ per flight $= 10^{-9}$ per hour.

Finally any improvement to the reliability of the standby system when active would be proportionately reflected through the contribution of case B_2 to the total risk.

APPENDIX 5-2
ASSESSMENT OF ALLOWABLE DEFICIENCIES

The purpose of this Appendix is to show the use of safety assessment in determining whether deficiencies in a system can be regarded as allowable.

Take a system similar to that in Appendix 5-1, except for the warning system which in this case is duplicated. Also, it will be supposed that all the systems are capable of being checked for operability before each flight begins. Hence dormant failures are confined to those which occur during the particular flight.

The numbers assumed are:–

$$T = 2 \text{ hours}$$
$$p_m = 10^{-4} \text{ per hour}$$
$$p_{sa} = 10^{-3} \text{ per hour}$$
$$p_{si} = 10^{-5} \text{ per hour}$$
$$p_w = 10^{-4} \text{ per hour}$$

Consider first the system free from deficiencies at the start of flight.

It is evident that if the main system operates throughout there is no risk. It is only necessary to consider main system failures in combination with other failures.

As before there are two conditions which need to be considered:–

Case A. Failure of the main system with dormant failure of the warning system.

Case B. Failure of the main system with failure of the standby:–

Case B_1 Dormant failure of the standby before main system failure.

Case B_2 Active failure of the standby after main system failure.

Case A

In case A both channels of the warning have to fail in the same flight as the main system failure. As there are only two hazardous fault sequences:–

A_1. Failure of warning channel 1;
Failure of warning channel 2;
Failure of main system.

A_2. Failure of warning channel 2;
Failure of warning channel 1;
Failure of main system.

i.e. $2 \times p_m p_w p_w \dfrac{T^3}{6} = 2 \times 10^{-4} \times 10^{-4} \times 10^{-4} \times \dfrac{8}{6} = 2 \cdot 67 \times 10^{-12}$

which is negligibly small.

Case B

In case B_1, the probability of this sequence of failures is:–

$$p_{si}p_m \frac{T^2}{2} = 10^{-5} \times 10^{-4} \times \frac{4}{2} = 2 \times 10^{-9} \text{ per flight.}$$

In case B_2, the probability is 2×10^{-7} per flight.

The total risk is, therefore:–

$$B_1 + B_2 = 2 \times 10^{-9} + 2 \times 10^{-7} = 2 \cdot 02 \times 10^{-7} \text{ per flight.}$$

We will assume that this represents a satisfactory level of safety for the day-to-day use of the particular system.

The question is then whether on occasions when one element of the system is deficient it would be safe to despatch the flight. Consider the possible deficiencies in turn.

With the main system inoperative, the flight could be made, on the standby only, with a risk of failure:–

$$p_{sa}T = 2 \times 10^{-3}.$$

With an inoperative standby, the flight could be made on the main system only, with a risk of failure:–

$$p_m T = 2 \times 10^{-4}.$$

With one channel of the warning system inoperative, the possible failures are:–

A.   Main system fail and second channel of warning fail.
B.   Main system fail and standby fail.

The probability of B is as above for the sum of B_1 and B_2.

As regards A the risk arises if the second channel of the warning fails before main system failure, with a probability,

$$p_m p_w \frac{T^2}{2} = 10^{-4} \times 10^{-4} \times \frac{4}{2} = 2 \times 10^{-8} \text{ per flight.}$$

Hence the total risk is:–

$$p_m \frac{T^2}{2}(p_{si} + p_{sa}) + p_m \frac{T^2}{2}(p_w) = 2 \cdot 02 \times 10^{-7} + 2 \times 10^{-8} = 2 \cdot 22 \times 10^{-7} \text{ per flight.}$$

From this examination it is evident that if either the main system or the standby has failed before flight begins the risk increases enormously, and such failures could not be regarded as allowable deficiencies. On the other hand, starting a flight with one of the duplicated warning channels known to be unserviceable only increases the risk about 10% above the normal level, provided that the other channel has been checked before flight and is serviceable. This might be counted a reasonable added risk for the occasional flight.

6
CASCADE AND COMMON-MODE FAILURES

INTRODUCTION

As was discussed in Chapter 1, the high levels of safety needed from essential systems are usually achieved by some form of "fail-safe" design, mainly by redundancy. In the case of structural members of systems the solution usually takes the form of alternative load paths or the use of crack stoppers. With electrical and hydraulic supply systems there are usually three or four channels powered by the main engines; these channels are sometimes supplemented by some form of standby system. With essential instruments there is usually duplication supplemented by some form of standby system. The degree of redundancy is determined not only by the required level of safety but also by the need to avoid delays to, and cancellations of, flights being caused by single failures in systems, so that some of the redundancy is often provided to cater for such cases.

In spite of these precautions, there are various threats to the independence of the channels of redundant systems which may lead to multiple failures at higher rates than would be forecast by calculating the multiple failure rates from the failure rates of the component channels. For example Fig. 6-1 shows the theoretical and practical curves for the probability of total failure of a system plotted against the number of channels available. The data is based on an analysis of the multiple channel failures which have actually occurred in a three-channel electrical system currently in service on a turbo-jet aircraft.

On the basis of the measured single channel failure rate of approximately 9.5×10^{-4} per flight, it might be expected that for a three-channel system the probability of all channels failing in one flight would be,

$$(9.5 \times 10^{-4})^3 = 8.6 \times 10^{-10} \text{ per flight}$$

whereas in practice it was several hundred times greater than this.

If one extrapolates these curves it will be seen that while the theoretical curve reaches extremely low values the practical curve tends to be asymptotic to about 10^{-7} per flight.

The cause of these apparent anomalies is that, for various reasons which are discussed below, the channel failures are not always independent. It is only by eliminating or reducing the effect of this lack of independence that proper advantage can be gained from the redundancy provided.

CASCADE FAILURES

In multi-channel systems the total systems loads are usually shared by the channels, so that in the event of failure of one channel its load or part of it will be shared between the remaining "healthy" channels. This increase in load is likely to produce some increase in the failure rate of the remaining channels. The example given above is an electrical one, but the same sort of increase in failure rate with increased load is likely to occur with other

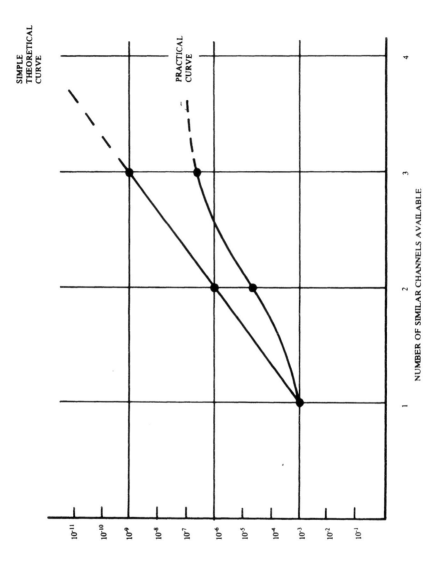

Fig. 6-1   PROBABILITY OF TOTAL FAILURE VS
NUMBER OF CHANNELS
(Courtesy of British Aerospace)

systems, such as, for example, a flying control system in which the surface hinge moment is shared between two or more hydraulic jacks.

An increase of failure rate of multi-channel systems resulting from additional loading can have a marked effect on the risk of combined failures. For instance, consider a system with identical duplicated channels each having a failure rate of 1 in 1,000 hours, in an aircraft with a mean flight duration of 1 hour. If both channels work continuously and share the load equally, the probability that one or the other channel will suffer a 'first' failure is $2 \times 1/1,000$. So in a period of one million hours, there would be about 2,000 such failures.

The probability of both channels failing in a flight is the product of the probability of one channel suffering a "first" failure and the probability of the other channel then failing. If the added load on the second channel does **not** affect its failure probability, then the combined failure probability is $(1/1,000)^2$; i.e. one double failure in one million hours. Suppose however that the effect of the added load after a first failure was to cause a drastic ten-fold increase of risk of a second failure, the combined probability is then $(1/1,000) \times (1/100)$; i.e. 10 such failures in one million hours.

Similarly, for a three-channel system for which the predicted failure rate could be 1 in 10^9 hours the achieved failure rate could be significantly worse than 1 in 10^7 hours, since there could be a very large increase in loading of the third channel following failure of the other two.

It is evident that the combined failure rates increase proportionally with increase of risk under the added load, and hence, it is important to take this into account. It is preferable to design to ensure that the channels cope with the added load without materially worsening the failure rate.

It is often the practice to base failure probability on the mean time between failure (MTBF) experienced in service. Thus, an MTBF of 1,000 hours per channel is taken as a failure rate of 1 per 1,000 hours. It should be noted that this is a good approximation to the failure rate under normal loading conditions, but gives little or no indication of the failure rate under added load. The reason can be seen from the above numerical example. In the running period of one million hours, 2,000 'first' failures occur, but only 1 or 10 (according to the effect of added load) double failures occur. Hence, the conventionally calculated MTBF is dominated by the incidence of first failures. Thus, while the MTBF derived from service experience is a useful guide to the normal failure rate, additional information is usually necessary to enable a prediction of the failure rate under extra load to be made. Attempts have been made to quantify the relationship between load and failure rate for some electrical components; the results of the work being detailed in a document MIL-HDBK-217B, "Reliability Prediction of Electronic Equipment" (Ref. 1). Similar work has been done by manufacturers of variable-speed accessory drives.

In considering likely failure sequences, one has to take account of the fact that following a series of failures the pilot himself will be under increased stress and may be more likely to make mistakes. There are recorded cases where after failure of two out of three electrical generators the flight-crew have mistakenly disconnected the remaining generator while carrying out emergency procedures.

Other types of cascade failure occur when a failure in a system or some part of the aircraft produces failures in other systems. The initial failure may be of a minor nature (e.g. a deflated tyre) but the overall consequences can be hazardous. Several examples of such failures were given in Chapter 2. Other examples include the following:—

a. The overcharging of an electrical storage battery, which caused it to catch fire and subsequently to cause a multiple failure of light alloy flying control rods in its vicinity; fortunately this occurred on the ground.

b. The failure of clamps joining sections of an air supply duct, causing leakage of hot air which in turn caused multiple failures of essential electrical circuits.

c. The failure of a joint in a domestic water system, which caused multiple failures of important electrical circuits. Similar multiple failures have been caused by leakage from galleys and by ingress of rain water through open doors or hatches on the ground.

d. The ejection of debris from engines, causing damage to other aircraft systems.

e. A repeatedly applied short circuit in a lamp holder, supplied from a warning circuit which was part of a printed circuit, caused progressive damage to the printed circuit board in such a way that an associated monitoring circuit was made ineffective. The damage then spread to the main control circuit, so as to cause a runaway of a system which had been deprived of its monitoring and warning system.

f. A fire caused by a high resistance electrical joint in its turn caused total failure of an electrical system because all of the alternator control circuits were located above the location of the fire.

SINGLE ELEMENT FAILURES PRODUCING MULTIPLE CHANNEL FAILURES

Although most systems employ redundancy techniques, it will be found on examination that many of them have a 'single element' or 'common point', the failure of which will cause multiple channel failures. Some are obvious to the analyst, some are not so obvious. The following are examples in various systems.

Electrical Systems

An obvious example of a common-mode failure is a system in which several generators are connected to a common bus-bar and where load sharing control is achieved by interconnections between the generator control equipment. In the oversimplified example shown in Fig. 6-2, it might be presumed that if the failure rate of each of the three generators with its control gear was 1 in 1,000 hours, the failure rate of the whole system would be 1 in 10^9 hours if the average flight time was 1 hour. However, it is unlikely that the failure rate of the bus-bar and the control system interconnections would be better than 1 in 10^6 hours, so that this would be the best failure rate obtainable from the system irrespective of how many generators were used.

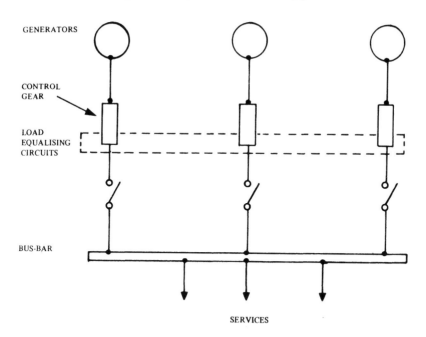

Fig. 6-2   THREE-GENERATOR SYSTEM WITH COMMON BUS-BAR

Multiple failures can also be caused as the result of using common earth points. The open circuit failure of a common earth point will cause the loss of all of the services connected to it, regardless of the fact that they may have separate electrical supplies. Not so obviously, a high resistance joint in the earth connection in Fig. 6-3 will cause an a.c. component to be injected into the d.c. equipment and a d.c. component to be injected into the a.c. equipment with possible malfunctions, particularly where sensing and amplifying circuits are involved.

In addition to causing failures or malfunctions of redundant channels, there is also the possibility of a fault which not only causes the malfunction of a system but also inhibits the device which monitors the system and disconnects it or gives warning of its failure or malfunction. Fig. 6-4 shows diagrammatically a control system (e.g. an auto-pilot) in which there is a monitor, which is a replica of the main control channel, and the output of which is compared with that of the main control channel, so that when there is a discrepancy above a pre-set value the output of the system is disconnected. The integrity of such a system can be undermined in any one of the following ways:–

a. Lack of adequate electrical and mechanical segregation between the main control and the monitor.

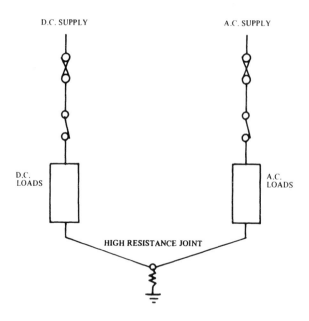

Fig. 6-3   COMMON EARTH FOR D.C. AND A.C. EQUIPMENT

b. Lack of adequate segregation within the comparator leading to cross connection.

c. The use of common power supplies, so that power supply modulation or ripples or pulses can cause malfunction of the main control and at the same time inhibit the monitor.

d. Electromagnetic interference affecting both the main control and the monitor in the same way.

With circuits which use integrating and differentiating functions or other processing which may be sensitive to changes in time constants, it is often difficult to predict the precise effects of failures, so that it is essential that the effects of the failures are evaluated on a test rig or on the aircraft. Even so their effects may vary with manufacturing tolerances and ambient conditions so that careful planning of the tests is necessary. For example, an unchecked runaway of a pitch trim system (fortunately on the ground) was caused by a failure in an arm of a rectifier circuit in a secondary power supply which was common to the control and monitoring circuits. The resulting pulses caused an integrating circuit to generate a false demand signal causing the trim to runaway and at the same time affected the functioning of the monitoring circuit. This effect was only reproduced on the bench after repeated attempts.

The routeing of electrical cables is another important factor affecting common-mode failures in electrical systems. A cable loom can contain several hundred electrical cables for various duties and current carrying

78

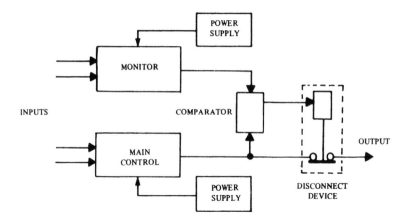

Fig. 6-4   TYPICAL CONTROL AND MONITOR SYSTEM

capacities, so that the failures to be considered are very numerous and not readily capable of assessment. For example multiple failures can be caused in the following ways:–

a. Serious overheating of one of the cables in the loom, because of inadequate electrical protection, causing multiple damage to adjacent cables.

b. Multiple damage to cables in the loom, because of lack of adequate mechanical protection (e.g. chafing and excessive mechanical load because of lack of support).

c. Multiple damage because of local fire or overheating in the region of the cables (e.g. because of leakage from hot air ducts).

d. Damage caused by mechanical failures in other systems (e.g. uncontained debris from engines or other rotating machinery).

e. Electromagnetic interference between electrical circuits.

Electrical plugs and sockets and junction boxes can be equally vulnerable to multiple damage.

It will be apparent that these types of common-mode failure can only be avoided by very closely controlled segregation of circuits, and in critical cases by special protection (e.g. armouring of cables, the use of fire-resistant cables). Special forms of protection are particularly appropriate to 'last-ditch' emergency systems which only come into action when a total system has failed because of some unpredicted set of circumstances.

Logic circuits which, while they may be multiplicated in each channel of a system, may each be accepting information from a common outside source (e.g. flap or undercarriage position sensors or air-speed switches). In addition they may be sensitive to the sequences of input signals so that there may be numerous combinations of failures and sequences of events which need examination.

Hydraulic Systems

Since hydraulic systems often provide the motive power for main and auxiliary flying control systems they may be the most critical systems in the aircraft. Normally, hydraulic power is generated by pumps driven from the main engines with back-up generation systems driven electrically or pneumatically or by ram-air turbines. To obtain the required reliability there are usually at least three systems, each being supplied by two or more pumps, which are so distributed between main engines as to ensure that after the loss of a single engine adequate power is left in each hydraulic system, and in the case of the multiple loss of engines, sufficient hydraulic power is left to control the aircraft at least when operating within a reduced flight envelope. However, in spite of these precautions common-mode failures can occur for the following reasons:–

a. The topping-up of each of the systems with the wrong fluid or with contaminated fluid.

b. The increase in load resulting from the loss of one system, causing increased failure rate in other systems.

c. Multiple damage to pressure and return lines, caused by debris from tyres, wheels, or engines.

d. The assembly of components such as non-return valves in a reverse sense.

e. Priority valves are often used in pressure and return lines to effect an automatic change-over in the case of a failure of the main supply system. However, if the failure is caused by leakage in the utilisation system or by a failure of the priority valve, the second supply will be lost as well. Also, unless non-return valves are judiciously situated, pipe, union or seal failures may also cause multiple losses. Similar multiple failures can be caused by other forms of change-over valves.

f. With jacks operating in parallel onto a common control surface (see Fig. 6-5) it may be possible for the malassembly or jamming of a valve to cause a hydraulic lock in one jack which cannot be overpowered by the other jack(s) in parallel, unless special design precautions are taken.

g. With tandem jacks there is a temptation, for reasons of simplicity and weight, to run hydraulic passages connected to several supply systems through forgings or other parts. While it can sometimes be shown that the forgings have safe fatigue lives or have adequate crack propagation properties, one can still be left with problems arising from common-mode failures. For example, the safe-life concepts can be undermined by undetected faults occurring in forgings, inaccuracies or damage in drilled passages or lapses in detailed design, such as was illustrated in Fig. 2-2, or in other secondary attachments to the main forgings (e.g. change-over valves). The safest course is to assume that within a common forging or casting etc., all hydraulic supply can be lost.

h. Unpredicted levels of vibration in particular locations can result in very high failure rates of pipes and unions.

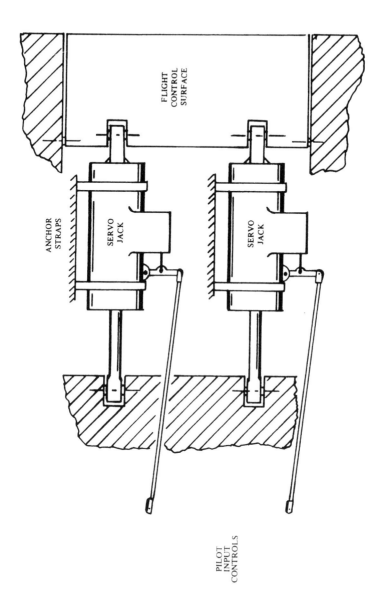

FLIGHT CONTROL SURFACE

ANCHOR STRAPS

SERVO JACK

SERVO JACK

PILOT INPUT CONTROLS

Fig. 6-5   SCHEMATIC EXAMPLE OF FLIGHT CONTROL
SURFACE OPERATION

(Courtesy of British Aerospace)

81

Mechanical Linkages

Most systems utilise mechanical linkages to transmit motion. These can be affected by two main types of failure, namely disconnection or jamming or combinations of these.

Hazards which can be caused by disconnection are usually overcome either by duplication of control rods, linkages etc., or by the provision of alternative methods of control (e.g. the differential use of spoilers as back-up for ailerons, or the use of pitch-trim devices independent of the main elevator control). Where safety is critically dependent on duplication of mechanical control runs, a search for possible causes of common-mode failures is important. The following are some possibilities of common-mode failures of mechanical linkages.

a. If the two parts are in close contact, as they may be in the case, for example, of a duplicated lever or shaft, they may fret against each other, or corrosion may occur so as to affect both of them, or they may both be damaged in the same way e.g. by someone standing on them.

b. If the two parts share the load equally and the failure of one is caused by fatigue, there is the possibility that the other part is nearing the end of its fatigue life, so that it may fail under the increased load. This problem is dealt with more fully, when discussing structural aspects, in Chapter 8.

c. As with electrical and hydraulic systems, both parts, if in the same vicinity, may be damaged by the same event (e.g. tyre fragments).

d. If one part is contained within another (e.g. a rod within a tube) common corrosion may occur and in addition the failure of the inner member may be difficult to detect.

Jamming or seizing can cause total failure of duplicated control runs. There are various causes including the following:–

a. The accumulation of frost, particularly when control rods and cables pass through pressure bulkheads where there is leakage of warm moist air from the cabin into a cold enclosure.

b. The accumulation of frost on unprotected screw jacks.

c. The presence of foreign objects (e.g. loose tools, nuts and bolts), drooping cable looms, loose panels etc., in the vicinity of control runs. This is particularly so where the control runs are close to horizontal surfaces liable to act as catchment areas, or where the ends of levers work in fairings or sump-like depressions, or where cables pass through holes or slots into which foreign objects can fall.

d. The working loose of a bolt, pivot etc., in such a way that it can foul adjacent parts having relative motion.

e. When a linkage breaks or becomes disconnected it may drop down and foul a static part and thus prevent movement of the unbroken linkage.

f. The jamming of a hydraulic valve, caused, for example, by distortion or by the ingress of foreign material.

Protection against jamming of duplicated control runs is primarily provided by careful design and also by the use of such devices as spring struts and disconnect devices, so situated in the control runs that should a jam occur in one control run the other path can still be operated. The jamming of hydraulic valves can similarly be protected against by the use of collapsing springs which allow the relative motion of valves to cut off hydraulic pressure. One design of jack uses concentric spool valves which normally move in unison but which in the case of one part jamming allow the other part to maintain some measure of control. (Ref. 5).

Fuel Systems and Multiple Engine Failures

Fuel systems include the storage, measurement, delivery, and control of fuel used in engines and the refuelling and jettisoning of this fuel. The airworthiness requirements for these systems stipulate that in at least one configuration of a system, and this applies particularly to take-off, the fuel supplies to each engine shall be completely independent. This effectively means the provision of at least one fuel tank and delivery system for each engine with isolating valves between the individual systems. Interconnection between the systems is used in cruising flight to achieve optimum distribution of fuel. The use of complex electrical systems to control the fuel delivery to turbine engines can introduce threats by common-mode failures to the independence of these engines, particularly from the aspect of electrical power supply, electromagnetic interference and errors in the logic and programming of the systems.

Multiple engine failures or malfunctions have been caused by refuelling with contaminated fuel and the filling of water methanol systems with fuel. In addition, multiple failures have been caused by icing-up of fuel vents and by pilot error in operating fuel selector valves. Also, on twin-engined aircraft, several accidents have been caused by the pilot shutting down the wrong engine after the misbehaviour of one engine.

Since electrical and hydraulic supply systems are usually powered by generators and pumps driven by the propulsion engines, the total shut-down of these engines may not only lose propulsion but also electrical and hydraulic supplies and, consequently, the control of an aircraft which may be dependent upon these supplies; additionally, the loss of electrical power may mean the loss of services essential for restarting the engines.

Since the shut-down of all engines even on a four-engined aircraft is not unknown, it is essential, in the event, to keep such supplies working as will enable the crew to maintain control of the aircraft and attempt to restart the engines during the descent and, in the limit, to attempt a landing on whatever terrain is available. This is usually achieved by powering the emergency systems by an auxiliary power-unit, by lowering a ram-air turbine into the airstream to provide emergency electrical or hydraulic power, or by having sufficient windmilling power from the engines to provide emergency hydraulic supplies, plus sufficient reserve electrical battery supply to operate instruments and engine restarting equipment.

Equipment Cooling Systems

Most modern aircraft have cooling provisions for avionic systems, in which the avionic equipment is mounted in racks which incorporate plenum chambers through which cooling air flow is induced by pressure difference provided by leakage to the outside atmosphere, supplemented at low altitude by electrically driven fans.

Failures of air supply or leakage from ducts have caused multiple failures of equipment. In addition the modification of racks to incorporate more equipment can cause elevated temperatures, leading to an overall increase in failure rates of the equipment. It is, therefore, necessary to include a study of these cooling systems in the safety assessment.

Ref. 1 gives considerable information on the variation of failure rates with ambient temperature.

CAUSES ASSOCIATED WITH EXTERNAL EVENTS

The effects of icing have already been discussed. Other external events include the following.

Bird Strikes

The current Joint Airworthiness Requirements state:–

> The aeroplane must be designed to assure capability of continued safe flight and landing of the aircraft after impact with a 4 lb bird when the velocity of the aeroplane (relative to the bird along the aeroplane's flight path) is equal to V_C at sea level or 0.85 V_C at 8,600 ft whichever is the more critical . . .
> Consideration should be given in the early stages of the design to the installation of items in essential services, such as control system components and items, which if damaged, could cause a hazard, such as electrical equipment. As far as practicable such items should not be installed immediately behind areas liable to be struck by birds.

Compliance with this requirement may involve a detailed assessment of the protection and segregation of essential services.

Lightning and Electrostatic Effects

On world-wide average, large aircraft are struck by lightning about once in 6,000 hours. In Europe the rate is higher, about one strike in 2,400 hours, which is about once per year per aircraft, so this is not an uncommon occurrence. In the main the strikes occur between the extremities of the aircraft (wing-tips, nose and tail) but they can sweep along the fuselage or across the wing behind projections such as the engines.

The principal risks are the ignition of fuel vapours at vents and the disruption of non-metallic unbonded parts.

There have been recorded cases of total temporary disconnection of electrical generating systems and several multiple failures of radio equipment, although nowadays effective aerial protection systems exist.

In spite of the Faraday cage effect of the metal skin, the high discharge currents through the skin, the associated voltage differences and the dynamic coupling can cause voltages to be injected into systems in the aircraft, particularly if these are earthed to the airframe. While this has not caused hazardous conditions in the past, there is the possibility with the extensive use of sensitive electrical circuits and digital systems that problems will arise. This has been the subject of research in several countries including the UK and it is anticipated that commonly agreed codes of practice to counter these lightning hazards will soon be published.

The Joint Airworthiness Requirements contain detailed regulations and advisory material concerned with the effects of lightning strikes (Ref. 2). See also Refs. 6 to 8.

Non-metallic external surfaces can build up static charges which can have undesirable and multiple local effects. For example, there have been total failures of duplicated electrical windscreen heating systems caused by coupling between the heating elements and surface static charges producing high voltages to earth on transformer secondary windings resulting in the breakdown of these. This was overcome by the use of 'leaky' capacitors connected between the circuits and earth. The use of carbon fibre composites for aircraft structure may give rise to problems such as local heating or induced currents in the vicinity of composite panels.

OTHER CAUSES OF COMMON-MODE FAILURES

Manufacture

A batch of faulty components can grossly alter the failure rate of a system and increase the probability of multiple failures. This type of failure is usually picked up by the operator and corrective action is taken, but until it is the system may run at a high risk.

The operator usually fixes 'alert levels' at which corrective action has to be taken. These alert levels are normally related to departures from the normal failure rates experienced in service. In addition failure rates should not be worse than those necessary to achieve the required level of safety (see Chapter 10).

Maintenance

The faulty setting up or rigging of equipment in multi-channel systems provides one of the most frequent causes of common-mode failures attributable to maintenance. Examples include:–

a. Errors arising from improperly calibrated test instruments or ground rigs.

b. The faulty rigging of sensors, giving switching signals which change the gain or mode of operation of critical systems. For example, an inadvertent stick-pusher operation was caused by the faulty rigging of duplicated micro-switches the function of which was to change the datum settings of the system relative to incidence.

c. The cross connection of parts of systems (e.g. electrical, mechanical, hydraulic) in such a way that it is not immediately obvious, but when the system is needed it behaves in a hazardous way, or that when a system fails the wrong action by the pilot is inevitable.

d. The incompleteness or incorrectness of maintenance instructions.

e. The interchange of equipment between different types of aircraft without making necessary adjustments such as those to sensitivity and gain. For example, during an accident investigation it was found that the ILS Localiser and Glide Path systems on a particular aircraft all had signal outputs set at a high level appropriate to another aircraft, so that not only did they make the course indicator over-sensitive but also inhibited the warning system.

The failure to re-establish proper restraint of electrical cable looms relative to moving parts such as control cables has not only produced serious electrical failures but also resulted in severance of the control cables. Also, the chafing of electrical cables on oxygen pipes has produced fires which have produced considerable multiple damage in crew compartments and elsewhere.

The designer can take various steps to keep these risks to a minimum, including:–

a. The use of detailed design precautions which minimise the risk of malassembly, poor rigging and cross-connections (e.g. the use of different end fittings in different channels).

b. Making critical areas readily inspectable.

c. Devising adequate check-out procedures to cater for maintenance errors which could result in hazards.

DEFENCES AGAINST CASCADE AND COMMON-MODE FAILURES

The prediction of rare events which are likely to produce these failures is almost, by definition, bound to be very difficult unless previous experience points the way. There are, however, various precautions and techniques which can reduce considerably the chances of such multiple failures. These include the following.

Segregation of Services

As will have been seen, a significant number of serious common-mode failures could have been avoided had greater attention been paid to mechanical and electrical segregation of components. After a design has been completed and failings found in segregation, it becomes very difficult to modify the offending features. It is, therefore, necessary in the early stages of design to identify clearly what the hazards are and what detailed techniques are to be used to avoid them and to inform those doing the detailed design accordingly.

In control systems it is often necessary to have cross-connections between channels in order to achieve synchronisation or load sharing or cross-monitoring. These cross-connections need considerable detailed attention and analysis supplemented by repeated testing of the effects of failures. Various techniques have been used for limiting the overall effects of failures, including the use of fibre-optics for transmitting information in order to avoid electrical and electromagnetic interference. Examples of the possible effects of interconnection in control circuits occurred in some early parallel d.c. generator systems where it was possible for a fault to elevate the voltage in one channel and take all the electrical load, making it appear that all of the other channels had failed. This frequently caused the crew to take the wrong action.

A modern wide-bodied aircraft provides greater opportunities for achieving physical segregation, particularly in the fuselage. However, problems have arisen on these aircraft, particularly in the region of the undercarriage bay and the engines because of the large amounts of energy which can be released because of tyre and wheel failures and engine disintegration and which are capable of causing damage over a large volume, so that to some extent the compensating effects of size are offset by the increase in available destructive energy and consequentially increased risk of multiple damage.

Various techniques of analysis are available to help resolution of these problems including Zonal Analysis and the study of trajectories in the case of uncontained failures. These are described below.

Zonal Analysis
In this form of analysis the aircraft is divided into zones and all of the equipment, cable runs, pipe runs etc., in each zone is listed. A study is then made of the effects of failures of this equipment on other systems within the zone and of threats from outside of the zone. If this analysis is not done until after the design is completed, any unacceptable findings may necessitate substantial redesign. It is, therefore, preferable to make this analysis as a continuous process during the design and to establish rules governing installation of components and systems to aviod common-mode failures. Table 6-1 overleaf shows a much simplified example of a Zonal Analysis of an underfloor area of the rear end of a cabin.

Non-containment of Fragments of Turbine Engines
Requirements for turbine engines ensure that minor failures, such as the shedding of one or more blades of the compressor or turbine, are contained within the engine carcase. However, weight considerations make it impractical to contain larger pieces of turbine and compressor rotors, and action has to be taken in the aircraft design to minimise the effects of the damage caused by uncontained debris.

The initial objective in framing the requirements was to limit the probability of a catastrophe from this cause to 1 in 10^8 hours. This probability can be regarded as the product of two component probabilities:–
i.e. Probability of non-containment of fragments
multiplied by
Probability of the fragments causing catastrophe.

Estimation of the first of these is an inexact matter, as it can only be determined by millions of hours of service experience, and this is not always available at the time a new aircraft is designed. For this reason the requirements concentrate attention on the second component of the probability with the object of limiting the risk of catastrophic consequences on the remote occasions when non-contained bursts occur.

TABLE 6-1
ZONE P – UNDERFLOOR, REAR END OF CABIN

Components in Zone P
Hydraulic pipe runs of systems 1, 2 and 3.
Elevator control cables.
Rudder control cables.
Tailplane trim electric cables.
Rudder trim cables.
Air ducts from rear engine to cabin ventilation system.
Air cycle refrigeration unit.
Rear engine control cables.
Electrical cables and connections from rear engine generators.
Electrical cables connections to rear engines.

Possible failures arising from components inside Zone P which could affect outside Zone P
Leakage of inflammable fluid from hydraulic pipes.
Leakage of hot air from air ducts.
Overheating of electrical cables.
Uncontained debris from air cycle refrigeration unit.

Possible failures arising from components outside Zone P which could affect Zone P
Uncontained debris from compressors of rear engines.
Decompression, caused by loss of underfloor hatches.
Leakage of fluids from toilet and galley compartments above zone.

Further, in order to keep the task within manageable proportions, the requirements reduce the infinite variety of possible sizes of fragments to two simple models, as illustrated in Fig. 6-6.

The first model is based on one-third of a disc. The trajectories of this fragment are assumed to be bounded to ±3° from the plane of the disc and to be spread uniformly round the full 360° viewed from the front. The risk of such a fragment causing catastrophe is limited to 1 in 20, this being the mean risk for all the turbine and compressor stages.

The second model simulates a piece of rim with blades attached with a mass assumed to be one-thirtieth of the bladed disc. The trajectories are assumed to spread to ± 5° from the plane of the disc, and the catastrophe risk is limited to 1 in 40.

Given these models, the first step in the safety assessment is to examine the 'geometry' of the situation. A plan view, Fig. 6-7, will indicate the systems and structure located in the damage swath of the fragments. A front view, Fig. 6-8, will show the radial spread of items in the damage paths. It becomes evident from diagrams of this kind that the portion of the full 360° occupied by vulnerable parts is considerably greater than the 1 in 20 or 1 in 40 permitted. Hence the need to proceed to the next stage of establishing whether damage will have catastrophic effects.

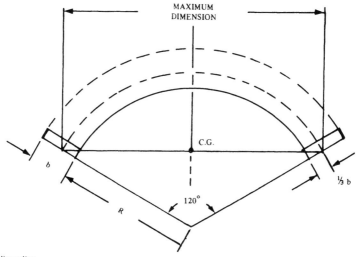

MAXIMUM
DIMENSION

C.G.

b

⅓ b

120°

R

R = disc radius

b = blade length

C.G. taken to lie on maximum dimension as shown

(a) SINGLE ONE THIRD DISC FRAGMENT

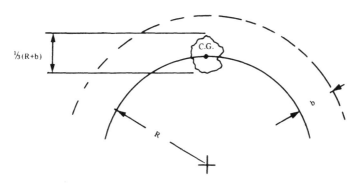

⅓(R+b)

C.G.

b

R

R = disc radius

b = blade length

C.G. taken to lie on disc rim

MASS ⅟₃₀th of bladed disc

(b) SMALL PIECE OF DEBRIS

Fig. 6-6   TURBINE ENGINE FRAGMENTS

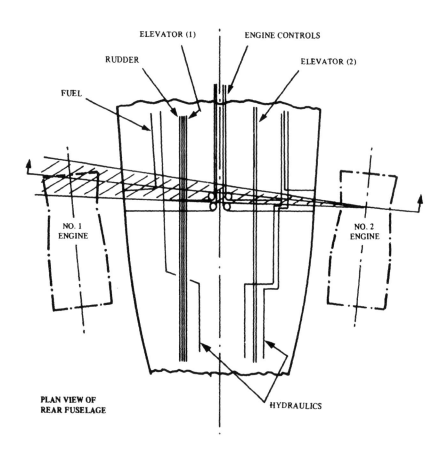

ELEVATOR (1)

ENGINE CONTROLS

RUDDER

ELEVATOR (2)

FUEL

NO. 1
ENGINE

NO. 2
ENGINE

PLAN VIEW OF
REAR FUSELAGE

HYDRAULICS

Fig. 6-7   TYPICAL FLY OFF ZONE OF ENGINE DEBRIS
(Courtesy of British Aerospace)

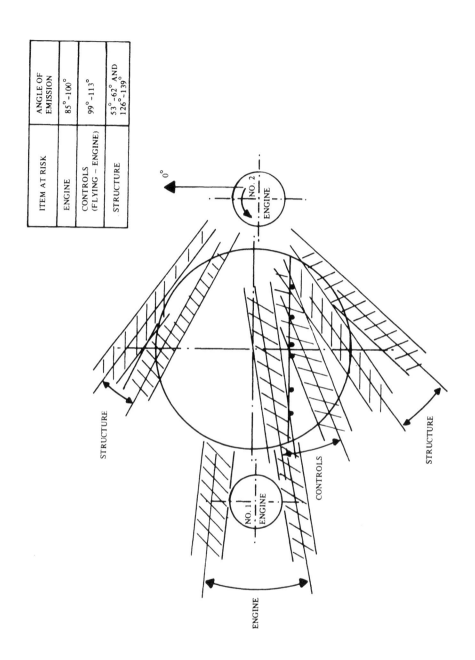

ITEM AT RISK	ANGLE OF EMISSION
ENGINE	85°–100°
CONTROLS (FLYING – ENGINE)	99°–113°
STRUCTURE	53°–62° AND 126°–139°

Fig. 6-8   ENGINE DEBRIS TRAJECTORIES (FRONT VIEW)
(Courtesy of British Aerospace)

In some cases, the consequences of damage will be quite plain. For instance, destruction of the entire elevator control system will be catastrophic, whereas destruction of one channel of a duplicated control system need not be. With multiplicated systems it is often possible by careful location to avoid all channels being destroyed by a single fragment. Thus, an important feature of the assessment is to guide the best layout of the design.

Fig. 6-9 illustrates the point. A and B represent two channels of a vital system, the failure of one channel being harmless, but of both catastrophic. A fragment from No. 2 engine just misses damaging both channels, but a fragment from No. 1 engine can, over a small angular range, damage both. Precise drawing would show the exact angular range. Suppose it were, say, 3°. The catastrophe risk, averaged over the two engines, would be

$$\frac{3}{360} \times \frac{1}{2} = \frac{1}{240}$$

Though this by itself is only a small fraction of the permitted 1 in 20, it is desirable to avoid it, if possible, by repositioning. If the parts were transferred to points X and Y, this would be achieved.

The energy of the large one-third fragment is so high that it is likely to penetrate all but the heaviest structure. Thus, the fuselage skin will offer no protection to the systems lying within. The lesser energy of the small piece of debris is such that it may be possible to establish that moderately strong structure, or specially designed armour, will serve to protect items lying behind it. Deflectors are possible to direct the fragment away from vulnerable items. However, these forms of safeguard need ballistic calculations, supported by tests, to justify their suitability.

In addition to the systems, attention is necessary to the consequences of damage to structure, adjacent engines, and fuel tanks. This leads to consideration of:–

a. Reduction of structural strength and stiffness.

b. Loss of thrust from the affected engine and adjacent engines damaged by debris.

c. Fire in tanks, or fire due to release of fuel from tanks or fuel lines, or increase of fire risk from damage to firewalls.

For these cases, considerable judgement is needed for the assessment of the severity of the effects of damage. For instance the reduced structural strength must be sufficient for the remainder of the flight. The requirements give guidance of the strength envelope which should be met.

Loss of thrust from one engine is unlikely to be of significance, but the loss of two engines can be serious depending on the number of engines in the aircraft, and on the phase of the flight during which the failure occurs.

The phase of the flight may also be of significance in other instances; for instance, the risk of fire which depends on temperature of the fuel. In cases where it is necessary to take account of the flight phase, the proportion of bursts which are likely to occur in the particular phase should be used. The

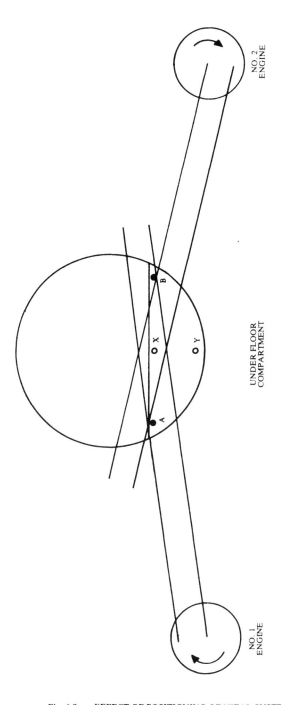

Fig. 6-9   EFFECT OF POSITIONING OF VITAL SYSTEMS

93

requirements quote statistical evidence of the proportion of all bursts divided between several defined flight phases. When a risk is present for a limited time, it is permissible to average the calculated risk over the whole flight.

The background of the JAR material and the methods used are more fully described in Refs. 3 and 4.

The Use of Dissimilar Redundancy

The use of dissimilar redundancy can be a powerful method of safeguarding against total failures of main systems. Various methods of achieving this are currently used on transport aircraft, either deliberately or in some cases fortuituously. The following are few examples of what has been done.

a. The attitude reference system is usually based on information transmitted electrically from electrically driven gyroscopes, so that essential artificial horizon instruments are entirely dependent upon the main electrical system. There have been several accidents which have been caused by the loss of attitude information in poor visibility conditions. Many aircraft are now fitted with a simple emergency artificial horizon instrument which in the event of failure of the main instruments is automatically switched to an electrical battery which is used only for this purpose.

b. Similarly, while the main heading information is dependent on electrical supplies it is backed up by a simple magnetic compass.

c. As has previously been stated engine-driven electrical generating systems are often backed up by drop-out ram-air turbine drives which are dedicated to the supply of essential services.

d. On some aircraft various supply services are used as back-up sources of power for other supply services. For example, on one aircraft the emergency electrical generator is driven by a hydraulic motor, on others emergency hydraulic pumps are driven by air-motors supplied with air from the engine compressors.

e. As described in Chapter 2, the VC10 aircraft, while using electric motors driving hydraulic pumps for the main flying controls, can still be handled in the case of total electrical failure by the use of tail trim and spoilers, powered by the main hydraulic system. In addition a drop-out ram-air turbine can be deployed to supply electricity in the event of a multiple engine failure. Other aircraft use electrical pitch trim while the main controls are hydraulically powered. Some aircraft have reversion to direct manual control in the case of system failure.

As will have become apparent, common-mode failures in multiplicated systems often occur because a single external event affects equally all the similar channels. The virtue of dissimilar redundancy is that because the channels are fundamentally different in their design, it is much less likely for an external event to affect them all in the same way.

The Identification of Critical Features

Whatever design precautions are taken to obviate common-mode failures, in the end there will be a residue of potentially hazardous situations which can only be made acceptable by special precautions in manufacture, inspection and maintenance and the way in which the aircraft is operated. It is, therefore, essential that these hazardous situations are properly identified so that the appropriate precautions can be taken and properly monitored.

References

1 MIL-HDBK-217B – Reliability Prediction of Electronic Equipment, Department of Defense, USA.
2 Joint Airworthiness Requirements, published for the Airworthiness Authorities Steering Committee by Civil Aviation Authority (UK).
3 Report CP-2017 – An Assessment of Technology for Turbojet Rotor Failures, August 1977, NASA (USA).
4 Paper AIR 1537 – Report on Aircraft Engine Containment, October 1977, Society of Automotive Engineers Inc., 400 Commonwealth Drive, Warrendale, P.A. 15096.
5 "How to Design a Safe Aircraft", George A. Peters, Hazard Prevention, (The Journal of the System Safety Society), March/April 1979.
6 Society of Automotive Engineers, Conference, "Lightning and Static Electricity", 1968.
7 Advisory Circular AC20-53 – Protection of Aircraft Fuel Systems against Lightning, FAA.
8 Culham Laboratory, Conference, "Lightning and Static Electricity", UK, 1975.

7
METHODS AND TECHNIQUES OF SAFETY ASSESSMENT

PLANNING THE ASSESSMENT
General
The content of a safety assessment will vary considerably according to such factors as the complexity of the system, how critical the system is to the safety of the aircraft, and what volume of experience is available on the type of system being used and what new methods and technologies are being introduced. So, before making a detailed safety assessment it is necessary to make a Preliminary Hazard Analysis of the system in order to determine the depth of assessment needed. It is then necessary to define the 'safety objectives' of the system before entering into detailed investigations.

Definition of the System
In order to establish meaningful safety objectives it is first necessary to establish a proper definition of each system. This definition should include:–
a. The intended functions of the system including its modes of operation.
b. The system performance parameters and their allowable limits (e.g. what departures constitute a failure).
c. The functional and physical boundaries of the system and of the components which comprise it.
d. The environmental conditions which the system will need to withstand.
e. The interfaces with other systems and with the crew.
f. Functional block diagrams of the system and its interfaces.

The definition should be in such terms that any change which goes beyond the original concept of the system (e.g. configuration, functions, environment) can be recognised.

Preliminary Hazard Analysis
Having established an adequate system definition the next step is to make a Preliminary Hazard Analysis. This should be concerned with the functions and vulnerabilities of the system, rather than with detailed analysis, and should define clearly what constitutes a failure condition of the system. It should include:–

a. The consequences to the aircraft and its occupants of the failure of the system or a part of the system to function within its specified performance limits.
b. The consequences of other possible malfunctions of the system (e.g. overheating) and their effects on other systems or parts of the aircraft.
c. The consequences to the system of failures in other systems or parts of the aircraft.
d. The identification of possible common-mode or cascade failures which may need detailed investigation.

e. The identification of possible sources of flight-crew error or maintenance error.

This Preliminary Hazard Analysis may well lead to modifications to the design in order to avoid some of the hazards or to mitigate their consequences. Having completed it the next step will be to decide which features of the system need detailed assessment, the extent of this assessment, and what test programmes are needed to confirm assumptions made in the analysis.

At this stage it may be decided that some features of the design are adequately covered by compliance with detailed airworthiness requirements for specific systems, and represent little change from systems already proven satisfactory or that the consequences of failure of the systems are not serious.

These Preliminary Hazard Analyses should produce an overall plan for the assessment of each system of the aircraft. This plan should be discussed and agreed with the Airworthiness Authority.

Safety Objectives

Having identified the hazards the next step is to formulate a set of 'safety objectives' appropriate to each defined system or function. These objectives should be consistent with the requirements defined in Chapter 4, but should be expressed in more detailed engineering terms appropriate to the particular functions and features of the system under consideration.

In some cases, such as the example below for an automatic-landing system, where failure rates, the incidence of atmospheric conditions and the relationship between specific failure conditions and the degree of hazard are reasonably predictable, it may be possible to express the safety objectives in purely numerical terms, as shown in Table 7-1.

In cases where there is a lack of reliable statistical evidence, but where the components are designed to standards known to produce satisfactory levels of reliability and have been subjected to extensive type testing and quality control, it should be possible to express the safety objectives in simple engineering terms.

For example, the following objectives could be applied to a hydraulic system on which the aircraft was entirely dependent for its main flying controls, and which consisted of pumps driven by main engines feeding the main systems and pumps driven separately from the main engines providing an emergency system for use in the case of multiple engine failures:–

a. The total failure of all hydraulic supplies shall be extremely improbable.

b. The aircraft shall remain controllable within the limits defined by X (a company document defining acceptable controllability in emergency situations) after the loss of any two hydraulic systems.

c. The aircraft shall remain controllable within the limits defined by X after the shut-down of all main engines.

d. The correct functioning of the emergency hydraulic system shall be capable of being checked in normal flight.

TABLE 7-1

SAFETY OBJECTIVES FOR A PARTICULAR AUTOMATIC-LANDING SYSTEM

Description of Failure Condition	Severity of Effect	Risk of Catastrophe	Objective
Touch down short of the runway	Hazardous	1/30	$<3 \times 10^{-8}$
Touch down with one wheel off the side of the runway	Hazardous	1/30	$<3 \times 10^{-8}$
Touch down on the runway but one wheel runs off the side of the runway	Hazardous	1/30	$<3 \times 10^{-8}$
Landing gear collapse	Hazardous	1/10	$<10^{-8}$
Wing tip touches ground before the wheels	Hazardous	1/10	$<10^{-8}$

NOTE: In this table the ratio between the number in the "objective" column and the required level of 10^{-9} in relation to a catastrophe, represents the estimated risk of the particular hazardous effect resulting in a catastrophe. For example, the risk of a catastrophe resulting from the hazardous condition "touch down with one wheel off the runway" is estimated at 1/30 based on a study of accidents on aircraft of a similar configuration (see Chapter 8).

These objectives would probably lead to a system comprising at least three independent, main engine-driven hydraulic systems, any one being capable of providing adequate controllability and each having a predicted failure rate of better than 10^{-3} per hour, supplemented by an emergency system having a failure rate of about the same order. These are conservative figures which are capable of confirmation, and almost certainly capable of being improved upon after a relatively short period in service.

However the objectives stated impose limitations on the scope of the analysis in that:–

a. Apart from the multiple shut-down of all engines they do not cover common-mode and cascade failures.

b. Apart from the checking of the emergency system, they do not cover 'dormant' failures (e.g. of selector or priority valves).

c. They only include multiple hydraulic failures. However, the system may well include electrical components, for example, electrically operated valves and warning systems, so that these too have to be taken into consideration.

It is better, where practical, to express the objectives both in numerical and qualitative terms, each one being a back-up for the other.

The above examples are very much simplified for the purpose of illustration. They need supplementing by guidelines in the form of acceptable practices (e.g. in relation to segregation) so that the detailed design can be checked as being consistent with the overall concept.

FAILURE ANALYSIS

Failure Condition

The requirements of JAR define a failure condition as follows:–

> A Failure Condition is defined at the level of each system by its effects on the functioning of that system. It is characterised by its effects on other systems and on the complete aircraft. All single Failure Conditions and combinations of failures, including failures in other systems, which have the same effects on the system under consideration are grouped in the same Failure Condition.

For example, in the simple lighting circuit illustrated in Fig. 7-1 there are numerous combinations of failures which can lead to the Failure Condition "loss of light".

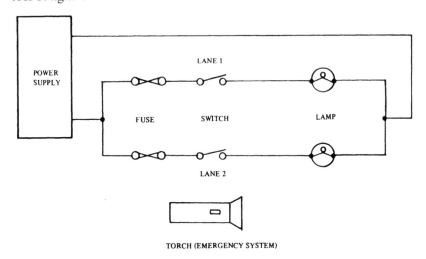

Fig. 7-1   SIMPLE LIGHTING SYSTEM

Failure Analysis

There are two main methods of making a failure analysis and these are well described in SAE Aerospace Recommended Practice ARP 926A (Ref. 1). The methods are sometimes referred to as "top-down" and "bottom-up".

The "top-down" or "functional" method starts by identifying the failure condition to be investigated, and then proceeds to derive those failure modes and combinations of failure modes which can produce the failure condition which is being investigated. The "bottom-up" or "hardware" method starts with the hardware failure modes which can occur, and analyses the effects of these on the system and the aircraft in order to determine the failure conditions which can occur. In both cases one studies the effects of the resultant failure conditions on the aircraft, and determines their seriousness and whether the safety objectives are fulfilled.

100

Whether the "functional" or the "hardware" method is used depends upon the nature and the complexity of the system. For example, if one starts with the "hardware" method for a complex system and needs to consider combinations of failures together with the effects of crew and maintenance errors, one may be faced with such an impossible number of combinations that one is driven to the "functional" method. On the other hand, if a module of a system proves to be particularly critical this module may have to be subjected to the "hardware" method, as may a critical mechanical device which may have a diversity of failure modes. A compromise between the two methods is usually adopted, with a "functional" approach being used to deal with combinations of failures, and a "hardware" approach being used to determine critical single failure modes.

The introduction of the examination of cascade and common-mode failures into the failure analysis is also a matter of a combined approach. One first determines those multiple failures which will hazard the system and then forecasts ways in which these multiple failures can be produced. However, it is also necessary to make a Zonal Analysis as described in Chapter 6. The prediction of multiple failures and their resultant "failure conditions" is thus, not only a matter of routine checking, but also of exercising considerable engineering judgement and having experience in this form of analysis, together with a wide knowledge of the features of the aircraft under review.

Fault Tree Analysis

The "fault tree" is a graphical method of expressing the logical relationship between a particular failure condition and the failures or other causes leading to the particular failure condition. The "fault tree method" is well described in ARP 926A (Ref. 1) so it will only be dealt with briefly here. The method employs logic symbols, as depicted in Fig. 7-2, in combination with other symbols representing events or cross references to other parts of the fault tree, as shown in Fig. 7-3.

A simple example of a fault tree analysis is given in Fig. 7-4, which depicts the analysis of the simple lighting circuit of Fig. 7-1, which has a standby consisting of a self-contained torch.

This form of presentation, although somewhat tedious to prepare, has the advantage that it forces a discipline on the analyst, and it gives a visual representation of sequences and combinations of failures. It is possible to break it down into a master diagram depicting "top events" leading to a failure condition and supporting diagrams giving detail of individual branches.

Dependence Diagrams

Another, and a rather less complicated method of depiction which has been adopted by some European aircraft constructors, is the dependence diagram. This is a block diagram in which each block defines, for example, a failure of a part of a system and the conditions related to it and, where needed, the estimated frequency of the occurrence described. The blocks

OUTPUT

INPUTS

The 'AND' gate describes the logical operation whereby the coexistence of all input events is required to produce the output event.

OUTPUT

INPUTS

The 'OR' gate defines a situation whereby the output event will exist if one or more of the input events exist.

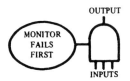

OUTPUT

MONITOR FAILS FIRST

INPUTS

The 'Priority AND' gate performs the same logic function as the 'AND' gate with the additional stipulation that sequence as well as coexistence is required.

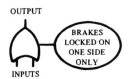

OUTPUT

BRAKES LOCKED ON ONE SIDE ONLY

INPUTS

The 'Exclusive OR' gate functions as an 'OR' gate with the restriction that specified inputs cannot coexist.

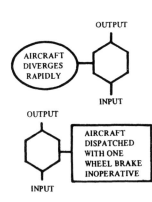

OUTPUT

AIRCRAFT DIVERGES RAPIDLY

INPUT

OUTPUT

AIRCRAFT DISPATCHED WITH ONE WHEEL BRAKE INOPERATIVE

INPUT

'INHIBIT' gates describe a causal relationship between one fault and another. The input event directly produces the output event if the indicated condition is satisfied. The conditional input defines a state of the system or a specific failure mode that permits the fault sequence to occur and may be either normal for the system or result from failures. It is represented by an oval if it describes a temporary condition that permits a fault sequence to occur; a rectangle is used to indicate a condition that is presumed to exist for the mission life of the system.

Fig. 7-2   FAULT TREE SYMBOLS – LOGIC OPERATIONS (GATES)

(Courtesy of Society of Automotive Engineers (Inc.))

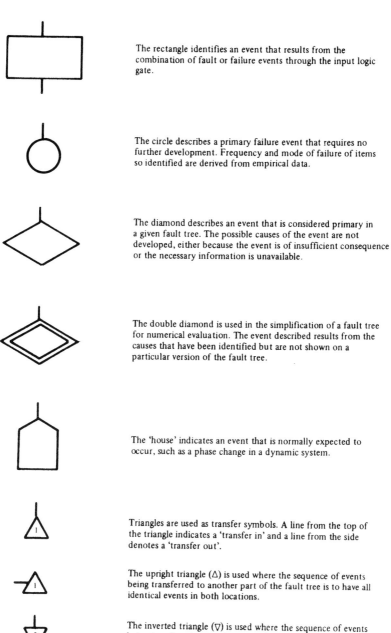

The rectangle identifies an event that results from the combination of fault or failure events through the input logic gate.

The circle describes a primary failure event that requires no further development. Frequency and mode of failure of items so identified are derived from empirical data.

The diamond describes an event that is considered primary in a given fault tree. The possible causes of the event are not developed, either because the event is of insufficient consequence or the necessary information is unavailable.

The double diamond is used in the simplification of a fault tree for numerical evaluation. The event described results from the causes that have been identified but are not shown on a particular version of the fault tree.

The 'house' indicates an event that is normally expected to occur, such as a phase change in a dynamic system.

Triangles are used as transfer symbols. A line from the top of the triangle indicates a 'transfer in' and a line from the side denotes a 'transfer out'.

The upright triangle (Δ) is used where the sequence of events being transferred to another part of the fault tree is to have all identical events in both locations.

The inverted triangle (∇) is used where the sequence of events being transferred to another part of the fault tree is to have one or more different events in the second location but is to be identical in function.

Fig. 7-3   FAULT TREE SYMBOLS – EVENT REPRESENTATIONS
(Courtesy of Society of Automotive Engineers (Inc.))

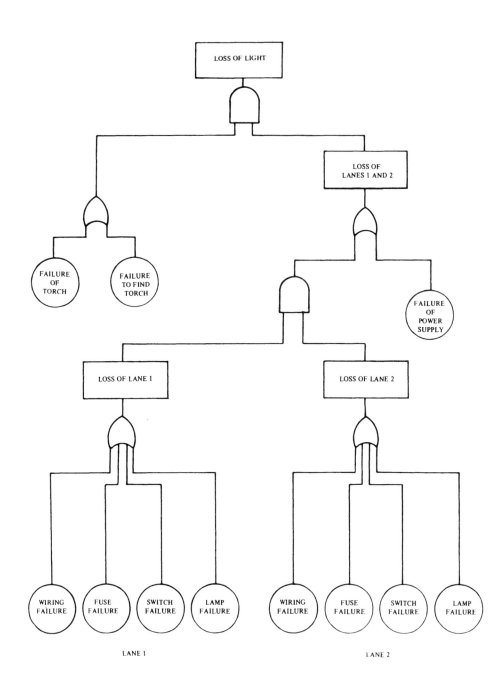

Fig. 7-4   FAULT TREE DIAGRAM OF FAILURE CONDITION
OF CIRCUIT OF FIG. 7-1

are arranged in series or parallel arrangements so as to represent "and" or "or" gates. Fig. 7-5 shows an example for the lighting circuit of Fig. 7-1. Diagram (a) of Fig. 7-5 represents the "top events"; the references in the bottom right-hand corners of each block refer to the sheet from which the information in that block was derived (as diagram (b)) of Fig. 7-5. This method of representation is particularly useful where numerical assessment of the probabilities of failure conditions are needed (Ref. 2). The probabilities appropriate to each box can be entered on the diagrams.

A further example is given in Fig. 7-6, which shows the top events in a particular failure condition of a droop-nose system on a supersonic aircraft. The system has main and standby actuation supplied by separate hydraulic systems and a mechanical free fall back-up. The failure condition is Major, since it will result in a high crew work load during landing, and so should be limited to a maximum of 1 in 10^{-5} per hour. Each block of this diagram will be supported by further dependence diagrams, and the probabilities appropriate to each block will be entered, so that the dependence for safety on the various parts of the system can readily be seen. The calculation of the overall probability of the failure condition depicted is of interest, and is discussed in Appendix 7-1.

Other Methods of Presentation
The fault tree analysis and dependence diagram are two methods of presentation which are commonly used in the USA and in Europe. There are other methods of representing failure conditions, including flow diagrams and columned forms of tabulation with specific entries. Individual firms will have preferences according to the nature of their projects and their experience. It is important that for a given project consistent methods should be adopted not only by the main contractors but also by sub-contractors. Otherwise, there is a danger of confusion and of mistakes being made if different methods of presentation are used.

THE USE OF PROBABILITIES
General
Having determined the failure conditions, one has to check whether they comply with the safety objectives, and since, in some cases, these objectives will be expressed in numerical terms, it will be necessary to evaluate the probability of occurrence of the failure conditions. Techniques for assessing these probabilities have already been described in Chapter 5 and an example is given in Appendix 7-1 to this Chapter, while techniques appropriate to performance evaluation are given in Chapter 8. The use of probability evaluations is particularly appropriate in the following instances:-

a. Checking whether the redundancy provided is adequate.

b. Determining how many failures to consider in any particular flight.

c. Determining check periods necessary to limit the presence of undetected (dormant) failures.

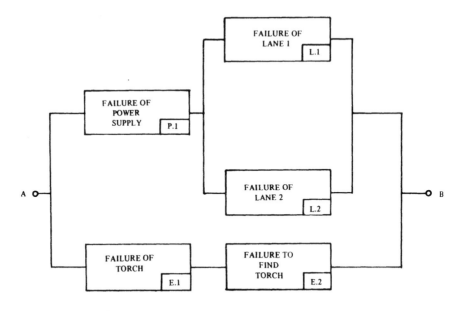

(a) TOP EVENTS LEADING TO TOTAL FAILURE

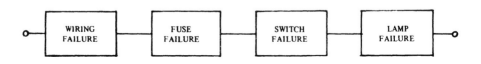

(b) DEVELOPMENT OF BLOCK L.1 (FAILURE OF LANE 1)

Fig. 7-5   DEPENDENCE DIAGRAMS OF FAILURE CONDITION
OF CIRCUIT OF FIG. 7-1

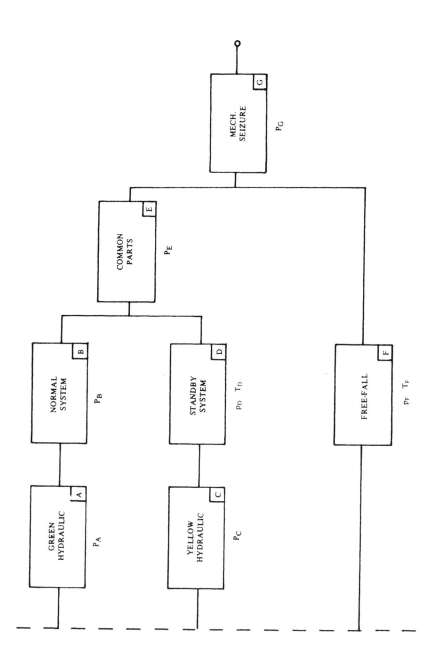

Fig. 7-6   NOSE DROOP SYSTEM – DEPENDENCE DIAGRAM
OF FAILURE CONDITION: FAILURE TO LOWER
BEFORE LANDING

107

d. Determining whether the effects of performance variations under normal and failure conditions are acceptable.

e. Determining what deficiencies of equipment are allowable before take-off (i.e. Master Minimum Equipment List – MMEL) and what restrictions should be applied if they are.

One can make probability assessments in a number of different ways including the following:–

a. By examining the modal failure rates of similar complete systems used in existing aircraft under similar conditions. This gives a feel for what should be achievable on the aircraft under similar conditions.

b. By examining the modal failure rates of similar components of systems used in existing aircraft under similar conditions.

c. By making piece parts analyses of particular components, using data such as that in document MIL-HDBK-217B (Ref. 3) for electronic components, or substantiated information from the manufacturer based on experience and testing.

Sensitivity

The aircraft constructor will use mixtures of these methods depending on the soundness of the information. There are problems, in that the forecast modal failure rates can vary substantially between equipment manufacturers. For example, at least an order of difference has been known between predictions for some pressurisation and hydraulic components; these differences are probably due to the lack of feed back from the operators and other overhaul organisations, and the fact that component "removal rates" are generally the only reliable data available rather than the confirmed failure rate. It is, therefore, necessary when using modal failure rates (see b. above) to be careful not to use over pessimistic or over-optimistic figures, particularly in the case of critical items. Where a high reliability figure is assumed in the case of a critical item, it is important to ensure that the design, the manufacturing processes, and the operating and maintenance instructions are consistent with obtaining this assumed reliability.

Having rationalised the information collected from component manufacturers and operators, it will be necessary, in the case of critical failure conditions, to examine how sensitive the predictions are to the effect of errors in the failure rates assumed for the dominant components.

For example Fig. 7-7(a) is a dependence diagram for a three-lane system each lane of which consists of three different components (A, B & C) the failure rate of each component being 2×10^{-4} per hour (i.e. an MTBF of 5,000 hours) so that the predicted total failure rate of the system will be,

$$(3 \times 2 \times 10^{-4})^3 = 2 \cdot 16 \times 10^{-10}.$$

Suppose that the failure rate of A is increased tenfold, then the total failure rate of the system will be,

$$[(2 \times 2 \times 10^{-4}) + (2 \times 10^{-3})]^3 = 1 \cdot 38 \times 10^{-8}$$

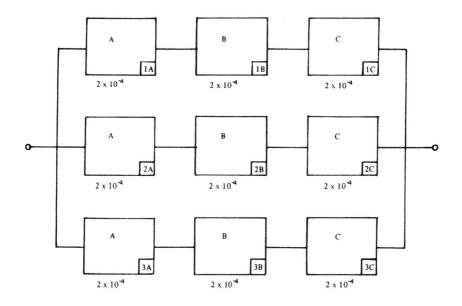

Fig. 7-7 (a) THREE IDENTICAL LANES

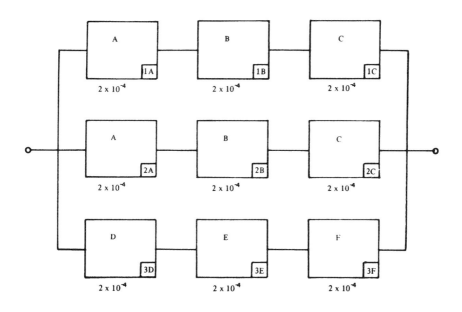

Fig. 7-7 (b) TWO IDENTICAL LANES AND ONE DISSIMILAR LANE

that is an increase of 64 times. If the third lane is altered so that it contains components D, E & F which are different from A, B and C (see Fig. 7-7(b)) each having a failure rate of 2×10^{-4} per hour, and the failure rate of A is again increased tenfold, then the total failure rate will be,

$$[(2 \times 2 \times 10^{-4}) + (2 \times 10^{-3})]^2 \times [3 \times 2 \times 10^{-4}] = 3 \cdot 456 \times 10^{-9}$$

that is an increase of 16 times. However, if the failure rate of item D is increased tenfold, the failure rate of the system will be $8 \cdot 64 \times 10^{-10}$, that is an increase of only four times.

Thus, there can be large differences in sensitivity to variations in individual failure rates and it is part of the designer's task either to reduce sensitivity by the arrangement of components or to ensure that sufficiently close control of the manufacture of the critical items is maintained.

Phase of Flight and Time of Exposure

The degree of hazard involved in particular failure conditions will depend on the phase of flight in which they occur (i.e. taxying, take-off, climb, cruise, descent, approach and landing), so that in making a safety assessment it is important to be clear of the effects of each failure condition in the appropriate phase of flight. In addition, some failure conditions may be more likely to occur during a particular phase of flight.

Because, in some cases, the use of the system is limited to a particular phase of flight, the time of exposure to risk may be limited. For example, with an automatic-landing system which has the capability of being checked before commencing the critical part of the approach, the time of exposure to failure conditions may be limited to a few minutes or even less. It is important in such cases that the calculations take account of any 'dormant' failures which may escape the checking system.

On the other hand, the fact that the system is only used for a short period, particularly if this is toward the end of the flight (e.g. flaps, brakes, undercarriage) may be a disadvantage in that failures which have occurred in earlier phases of flight may not have been detected and will only become apparent when the system is needed.

CREW AND MAINTENANCE PROCEDURES

Crew Warning and Crew Drill

In making a safety assessment account will have to be taken of warnings and indications given to the flight-crew and drills which are to be followed by them during normal operation and operation with failure conditions. In the case of some failures, it will be assumed that the crew will take action within given periods depending upon the urgency of the warning system or the behaviour of the aircraft.

It is, therefore, important that the Flight and Operations Manuals are consistent with the assumptions made in the assessment. In fact, many of the crew drills will be a direct outcome of the safety assessment. The assessment should, therefore, record the assumptions made regarding crew drills.

Pre-flight Checks and Maintenance Checks

The safety assessment will usually reveal that pre-flight checks of some critical features and routine maintenance checks at defined periods are essential in order to achieve the required level of safety. This is particularly so in the case of checks to reveal dormant failures.

The periodicity of these checks will be used in the probability calculations in the assessment, and should, therefore, be carefully recorded in relation to critical failure conditions, so that any alterations to these periods can be referred back to the appropriate part of the analysis.

THE ASSESSMENT OF VULNERABILITY TO ERROR

As was discussed in Chapter 2, a substantial proportion of accidents are attributable to errors in manufacture, maintenance and piloting. However, as has already been pointed out, many of these accidents could have been avoided by more intelligent design. In order to reduce the likelihood of such accidents, it is necessary to make an analysis of the design to determine those features of it which are vulnerable to error and to eliminate these as far as practical, or otherwise to ensure that instructions, checks, training etc., can be relied upon to safeguard against the predictable errors. There will always remain the unpredictable error. Additionally, it is not often practical to make any credible prediction of error in numerical terms.

It is important that, during the analysis of the possibilities and effects of error, there is intense co-operation between the designer and maintenance engineers and pilots, as appropriate. One has to consider errors in as systematic and analytical a way as the failure of components in order to make any impression on the problem and to make considerable use of past experience (Refs. 4 to 8). In order to gain increasing knowledge of the possibilities of crew and maintenance errors it is essential that incidents which occur in flight are accurately reported (see Chapter 10). In assessing pilot error possibilities, the use of a representative simulator is invaluable. The following are three areas which can involve pilot error and need particular attention:–

a. **Warning Systems.** Are the warning systems and other clues given to the pilot adequate for him to recognise the problem and take the correct action when an unusual and hazardous failure condition occurs? This is a particular problem when an event such as an engine break-up, a stalling situation or a major electrical failure etc., produces a multiplicity of warnings. Also, where the results of a failure condition demand immediate and difficult correction of the flight path, the pilot is likely to ignore the warnings associated with the source of the problem until the aircraft has been re-established on a steady course. It is, therefore, necessary to look at the problem not only on a system-by-system basis, but also to consider the situation following the big events, such as those resulting from some of the predictable cascade failures.

b. **Secondary Controls.** A large proportion of the system accidents attributed to pilot error have been concerned with the incorrect operation of controls by the crew, and these for the most part have involved lift, lift-dump and drag devices. To the authors it appears unacceptable, in the light of the evidence, to have secondary controls which if wrongly operated are bound to produce a catastrophe. It would seem reasonable to ensure that either:–

i. there is an automatic inhibition of the operation of controls in the phases of flight where improper operation would produce a catastrophe, or

ii. the improper operation of the controls will become immediately obvious to the pilot and allow him sufficient time to recover from the resulting situation.

c. **Presentation of Information to the Crew.** Numerous accidents have been caused by the misreading of the information presented to the pilot, either because of ambiguity or lack of clarity in the presentation, or some distraction causing the pilot to ignore important instruments (e.g. altimeter, airspeed indication, course selection). This subject is dealt with in greater detail in Refs. 4 to 8.

The following are typical examples of accidents, the first involving the misreading of an altimeter.

24 December 1958   Britannia   Christchurch, Hants.   (9 fatalities)

The accident was the result of the aircraft being flown into ground obscured by fog. This was caused by a failure on the part of both the Captain and the First Officer to establish the altitude of the aircraft before and during the final descent. The responsibility for the accident must rest with the Captain. The height presentation afforded by the type of three pointer altimeter fitted to the subject aircraft, was such that a higher degree of attention was required to interpret it accurately than is desirable in so vital an instrument. This when taken in conjunction with the nature of the flight on which the aircraft was engaged was a contributory factor.

Other accidents, where similar misreadings on turbo-jet aircraft were suspected to have occurred, led the CAA to issue an Airworthiness Notice requiring the fitment of an altimeter which had a counter (digital presentation) and a pointer on turbo-jet aircraft of over 12,500 lb (5700 kg) AUW. A typical counter pointer instrument is shown in Fig. 7-8.

Another accident which involved the distraction of the crew from the main instruments is as follows:–

29 December 1972   L-1011   Miami   (99 fatalities)

The National Transportation Safety Board determines that the probable cause of this accident was the failure of the flight crew to monitor the flight instruments during the final 4 minutes of flight, and to detect an unexpected descent soon enough to prevent impact with the ground. Preoccupation with the malfunction of the nose landing gear position indicating system distracted the crew's attention from the instruments and allowed the descent to go unnoticed.

Accidents such as those above and those involving navigational error, led the FAA, the CAA and other airworthiness authorities to require the fitment of Ground Proximity Warning Systems to large transport aircraft to alert the crew to imminent danger of collision with the ground. This is an example of an overriding warning system which takes account of several aspects of crew error.

Fig. 7-8   **TYPICAL COUNTER POINTER ALTIMETER**
(Courtesy of Smiths Industries)

Accidents have probably been caused, particularly on some smaller aircraft, by the presentation of airspeed information in more than one unit (knots, mph or kph) where the clutter of information and the use of two units has led to error.

References

1 ARP 926A – "Fault/Failure Analysis Procedure", SAE Aerospace Recommended Practice, Society of Automotive Engineers (USA), May, 1979.
2 MIL-STD-756B – Military Standard, "Reliability Prediction", Department of Defense, (USA).
3 MIL-HDBK-217B – Military Standardisation Handbook, "Reliability Prediction of Electronic Equipment", Department of Defense, USA.
4 International Air Transport Association (IATA), 20th Technical Conference, "Safety in Flight Operations", Istanbul, 1975.
5 Dutch Airline Pilots Association (VNV), Symposium, "Safety and Efficiency: the next 50 years", Den Haag, 1979.
6 British Airline Pilots Association (BALPA), Technical Symposium, "Outlook on Safety", London, 1972.
7 Royal Aeronautical Society, Symposium, "Flight Deck Environment and Pilot Workload", London, 1973.
8 Flight Safety Foundation (USA) (FSF), 32nd International Air Safety Seminar, "New Technology & Aviation Safety", London, 1979.
9 Securité des Systèmes – C. Lievens, Cepadues, Editions, Toulouse.

APPENDIX 7-1
AN EXAMPLE

Description of System

Fig. 7-6 shows the dependence diagram for a system designed to lower a droop-nose prior to landing. It comprises a hydraulic power supply, A, operating the normal system, B, with a second hydraulic supply, C, to operate the standby system, D. The standby is only brought into use in the event of failure of either the first hydraulic supply or the normal system. The two systems join together in common parts, E. In the event of combined failure of the two systems, or of the common parts, the free-fall system, F, is brought into use. The systems finally join in further common parts, subject to the risk of mechanical seizure, G.

The average flight duration is T hours. The probability of failure of each of the components A, B, C, E, and G, in a flight of T hours can be estimated from consideration of the probabilities throughout each phase of the flight. The hydraulic systems will of course require their own detailed assessments as they serve other systems in the aircraft. The probabilities for these five components, designated P_A, P_B, P_C, P_E, and P_G, are found to be invariant, i.e. they have constant probabilities per flight.

The normally inoperative standby system, D, is subject to dormant failure risk. The probability for this is a function of the rate of occurrence p_D per hour, and of the time since the last check, the interval between checks being T_D. The same applies to the free-fall system, F, its rate of failure being p_F and intervals between checks T_F.

The Nature of the Calculation

The object is to determine the probability of total failure of the system to lower the droop-nose, and then to find whether this risk per hour, averaged over a full cycle of flights (in this case T_F) exceeds an allowable level. Total failure to lower is regarded as a Major Effect with a maximum level of 10^{-5} per hour.

Examination of the dependence diagram shows that the probability of total failure is in the form,

$$[(A + B)(C + D) + E] F + G$$

The allocation of probabilities to the terms A, B, C, E, and G, is straightforward as they are invariant from flight to flight. However, the probability of dormant failures in D and F increases with time and needs special consideration. The above expression can be re-written in the form,

$$[(A + B) C + E] F + (A + B) DF + G$$

It will be seen that in this form, in the first term the variable probability, F, appears by itself, while in the second term, the two variables, D and F, appear as a product.

Consideration of Component F

As a simplifying assumption it will be supposed that $p_F T_F$ is sufficiently small to justify taking,

$$1 - e^{-p_F T_F} \simeq p_F T_F.$$

(This can be checked numerically).

At the end of the first flight, before which component F had been checked as operative, the probability that failure will have occurred is then $p_F T$. At the end of the last flight before the next check, the probability is $p_F T_F$. The average over all the flights is thus,

$$\frac{p_F}{2} \left(T + T_F \right)$$

The average, **per hour**, is,

$$\frac{p_F}{2} \left(1 + \frac{T_F}{T} \right)$$

If there are a large number of flights between checks, a close approximation is,

$$\frac{p_F T_F}{2T}$$

Consideration of Components D and F Together

Again it will be assumed that,

$$1 - e^{-p_D T_D} \simeq p_D T_D.$$

The probability of dormant failure of D rises from zero to $p_D T_D$ at the end of its check period, and then falls to zero. Meanwhile the probability of dormant failure of F, starting from zero, increases right up to the end of its check period. The combined probability is in the form,

$$p_D p_F t^2$$

The combined probability is illustrated below.

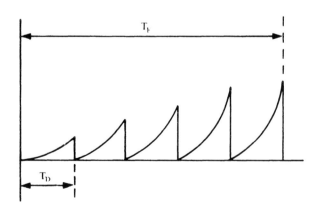

116

Consider the 'r'th period, when component D has reached $r - 1$ cycles. Counting time, t, as starting at the beginning of this 'r'th period, the probabilities of failure are,

Component D $\quad p_D t$.
Component F $\quad p_F [(r-1) T_D + t]$.

Over this 'r'th period, the combined probabilities average,

$$\frac{1}{T_D} \int_o^{T_D} p_D p_F t [(r-1) T_D + t] \, dt = \frac{p_D p_F}{6} T_D^2 (3r-1).$$

Over a complete set of periods, from $r = 1$ to $r = T_F/T_D$, the combined probability is,

$$\frac{p_D p_F T_D^2}{6} [2 + 5 + 8 \ldots .]$$

which is an arithmetical progression, the sum of which is,

$$\frac{p_D p_F T_D^2}{6} \left[2 + \left(\frac{3T_F}{T_D} - 1 \right) \right] \frac{T_F}{2T_D}$$

The average, **per hour**, over T_F/T_D periods is,

$$\frac{p_D p_F T_D^2}{12T} \left[1 + \frac{3T_F}{T_D} \right].$$

Total Probability

Having determined the values for the terms for F and for DF, these can now be entered in the original expression, giving the probability per hour as,

$$\left[(P_A + P_B) P_C + P_E \right] \frac{p_F T_F}{2T} + (P_A + P_B) \frac{p_D p_F T_D^2}{12T} \left(1 + \frac{3T_F}{T_D} \right) + \frac{P_G}{T}.$$

Suppose that the following numerical values apply,

$P_A = 18{\cdot}5 \times 10^{-4}$	$p_D = 0{\cdot}66 \times 10^{-4}$ per hour
$P_B = 2{\cdot}0 \times 10^{-4}$	$p_F = 0{\cdot}05 \times 10^{-4}$ per hour
$P_C = 4{\cdot}5 \times 10^{-4}$	$T = 3$ hours
$P_E = 0{\cdot}015 \times 10^{-4}$	$T_D = 3{,}000$ hours
$P_G = 0{\cdot}003 \times 10^{-4}$	$T_F = 6{,}000$ hours

The three terms in the above expression are then,

$$(1{\cdot}21 + 118{\cdot}3 + 10{\cdot}0) \times 10^{-8} = 0{\cdot}129 \times 10^{-5} \text{ per hour.}$$

This is well within the limit of 1×10^{-5} per hour.

Sensitivity

It is evident that in the middle term of the three dominates the overall probability. Thus, we can see at once that errors in estimating P_C, P_E and P_G would have little effect on the outcome. Also since P_A is large compared with P_B, errors in P_B would have a small effect. As a rough approximation, an error of 10% in P_A, p_D, or p_F, would alter the overall probability by 10%. Recalling that a simplifying assumption was made that,

$$1 - e^{-p_D T_D} = p_D T_D$$

this can now be checked numerically. The true value is $0\cdot1796$ and the approximation $0\cdot198$. Thus the calculation is somewhat pessimistic. For the $p_F T_F$ term, the true value is $0\cdot0296$ and the approximation $0\cdot03$, virtually the same numbers.

Changes of System

Having made this general calculation, it is then easy to find the effects of altering the system or varying the check periods T_D and T_F.

For instance we can find the results of omitting the standby system entirely. In this case, the overall probability is in the form,

$$(P_A + P_B + P_E)\frac{p_F T_F}{2T} + \frac{P_G}{T} = 1\cdot035 \times 10^{-5} \text{ per hour.}$$

This is marginally above the acceptable limit, but could be reduced to 1×10^{-5} by reducing the check period T_F to 5,800 hours. There may, of course, be practical reasons for preferring to retain the controlled operation of the standby system rather than relying on free-fall when the normal system is inoperative.

8
SOME PARTICULAR TECHNIQUES

THE ASSESSMENT OF PERFORMANCE OF SYSTEMS

Introduction

A lot of this book has been concerned with systems failures, their consequences and likelihood; but for some systems there will also be a need to examine performance from a safety viewpoint. In the main this applies to complex systems, usually automatic, where a number of sensing devices are coupled through computing elements to a control or an instrument. Although what follows will generally be referring to the performance of a whole system made up of a number of items of equipment, it may be just as applicable to an individual item of equipment.

Performance – What Is Meant By It?

In a nutshell, the performance of a system or equipment is the accuracy with which it performs its intended function. If all systems were built in precisely the same way, and operated under the same conditions, there would be no performance variability. However, in the real world there are many reasons for differences in performance, and these can be broken down somewhat arbitrarily into three main groups.

Firstly, there are those which directly affect the physical make-up of the system, such as manufacturing tolerances, maintenance adjustments and so on. These can lead to datum errors, variations in the response of the system to particular stimuli, etc.

Secondly, there are those things which indirectly affect the way the system responds in any given circumstances. These are largely environmental in that temperature, vibration, etc., are prime contributors but there may also be effects resulting from the characteristics of input supplies, such as voltages, hydraulic pressures.

Thirdly, there is the basic competence of the system in carrying out the job it is designed to do. This is largely a question of the ingenuity of the designer in his choice and treatment of the input parameters.

Performance and Safety – Why are Airworthiness Authorities Interested?

For many systems on an aircraft, performance is more a market-place item than an airworthiness safety item. For example, an auto-pilot which controls the aeroplane inaccurately may have a 'nuisance' value to the crew, or if it causes oscillations it may even make the passengers sick, but in either case there would be no hazard to the aeroplane. The bad performance is then a question of the guarantees to the operator from the manufacturer, and has to be resolved at a commercial level. There is only a need to consider poor performance in the safety assessment if it can be critical to safety, and if the cues for pilot detection are uncertain, or if there is insufficient time for him to react.

An obvious example is automatic landing where the flare-out is critical, and an inadequacy may only be detected by the pilot very late. This is further compounded by the fact that in most cases automatic-landing systems are designed for use in very low visibilities when the visual cross-checks from the outside world may be virtually non-existent. In such a case the safety assessment must establish that the probability of an incident due to performance is suitably remote. The remainder of this Chapter describes the measurement of performance and the assessment of its adequacy.

The Measurement of Performance

The 'output' of a system when carrying out the task delegated to it can be expressed as a statistical distribution which describes the probabilities that the system output will reach or exceed any particular values. The output depends on what the scope of the system is: for a radio navigation receiver it might be an electrical signal giving position information, whereas, for a whole automatic-landing system the touch-down position on the runway may be regarded as the output. In theory this distribution can be determined by measurement, i.e. if the system is operated a number of times and the output measured each time, a statistical distribution will result.

However, in a complex system which is affected by a number of variables, it may be difficult to ensure that a true description of the statistical distribution of system output can be achieved by this type of random testing. Firstly, a sufficient number of different (but nominally similar) units would have to be tested to ensure that the full range of system tolerances had been covered during the testing. Secondly, there needs to be confirmation that the full range of all environmental conditions has been encountered in their correct proportions.

An alternative to this form of random testing, is to break the performance distribution down into its components by considering the dominant variables affecting the output. Since each critical parameter will itself be subject to some form of statistical distribution, the testing should be structured in a way which ensures that the effects of all significant variables are correctly weighted in the final result. By testing the effects of individual variables separately (and in some cases together) and from a knowledge of their own distribution, the distribution of system output can be represented with reasonable confidence. Much of this type of testing can be carried out by simulation and bench-testing, whereas, if the overall performance distribution is to be based on random testing, then measurements of system output should be carried out for a larger number of real systems, in a representative spectrum of real conditions, probably involving many flight test points.

Example of Output Performance Distribution – Lateral Position in Auto-land.

The first step is to establish the acceptable level for the system performance, and then to determine that the contribution from the individual contributing factors is such that this level will be achieved.

Taking the case of lateral position during the landing, provided the wheels stay on the runway during the touch down and the landing roll then the landing can be judged to have been 'safe'. If, however, the wheels do leave the runway at any stage there is some probability that from this incident a Catastrophe will result. From a study of accidents, it has been accepted that a Catastrophe/Incident ratio of 1 in 30 can be used. Now, if it is accepted that a Catastrophe from this cause should have a probability of 10^{-9}, then the probability of the wheels being off the side of the runway should not exceed 3×10^{-8}. This is, effectively, the basic lateral performance requirement for the automatic-landing system.

However, it is impractical to leave the requirement in this form since part of the performance variability arises from the ground installation, and part from the aircraft and its equipment. Since these represent quite different spheres of interest, there has to be some apportionment of risk between the two areas.

If it is assumed that the overall distribution is Normal, then the number of standard deviations equivalent to 3×10^{-8} is 5·54. Runways have a half-width of 75 ft plus a 25 ft hard shoulder, and a typical aeroplane has a half-track of 15 ft, so that the aircraft centre-line can move ±85 ft about the runway centre-line before the wheels leave the hard surface. Thus one standard deviation should be 15·34 ft.

Studies have established that the variables listed in Fig. 8-1 are the significant ones and that a reasonable apportionment between them on a 'root mean square' basis (see Chapter 5) is as shown. The airborne side can be divided into two areas. Firstly, there is the accuracy of the ILS receiver in giving the position of the ILS beam centre-line, and then there is accuracy with which the auto-pilot (with all its other sensors) can fly to this reported centre-line. The accuracies given for ILS beams are reflected in the International Standards for such equipment (Ref. 1) while those relating to ILS receivers are contained in national procedures (Refs. 2 and 3).

ILS Receivers

In most countries ILS receivers are required to be approved by the Airworthiness Authority, and they are classified according to accuracy. In practice, this means that testing during the approval process must, amongst other things, establish that the desired level of accuracy is achieved over the whole range of normal operating conditions.

Minimum performance standards and test procedures for ILS localiser receivers which have been developed by the Radio Technical Commission for Aeronautics (RTCA) are published in Ref. 2. The list of parameters (random variables) which have been determined as significant is given in column 1 of Table 8-1 on page 123. The other columns give, for each variable, the probability that it will be encountered, its distribution, and the nature of its effect on centring error. With two exceptions the distributions are regarded as being Normal with a linear effect on centring error. The remaining two variables are regarded as having rectangular distributions within specified limits, and in one case (radio frequency CRF) level variation

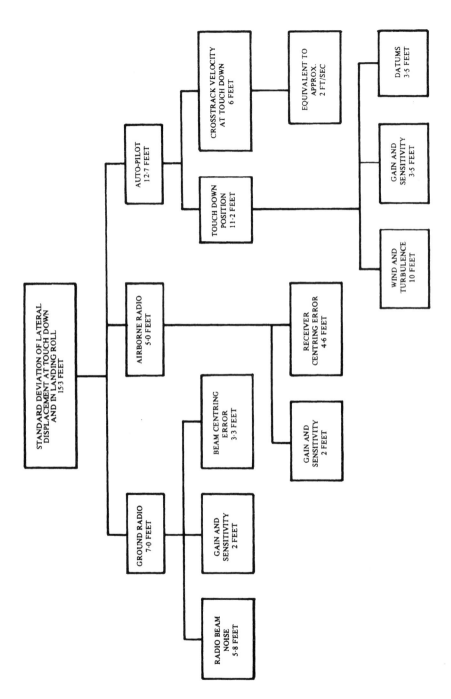

Fig. 8-1   CONTRIBUTION TO SCATTER IN LATERAL POSITION
DURING AUTOMATIC LANDING

the effect on centring is linear, while in the other case (temperature) the effect is less predictable. For all the variables with a linear effect, the contribution to the overall standard deviation of centring error can be established by two test measurements. For temperature, however, the effect has to be measured at intervals and then taken into account proportionately.

TABLE 8-1

STATISTICAL PROCEDURE FOR DETERMINATION OF
ILS RECEIVER CENTRING ERROR

ASSUMED DISTRIBUTION AND COURSE ERROR
FUNCTIONS OF PRIMARY VARIABLES*

Environment	Distribution of Primary Variable	Probability of Encountering Primary Variable	Error Function
1 RF Level Variation	Rectangular	1·00	Linear
2 Carrier Frequency Variation	Normal	1·00	Linear
3 Cross Modulation	Normal	0·05 (G.S.) 0·025 (LOC)	Linear
4 Power Source Frequency Variation	Normal	1·00	Linear
5 Power Source Voltage Variation	Normal	1·00	Linear
6 Modulation Frequency Variation	Normal	1·00	Linear
7 Modulation Phase Variation	Normal	1·00	Linear
8 Modulation Per cent Variation	Normal	1·00	Linear
9 Voice/Ident Modulation Frequency Variation	Normal	0·025	Linear
10 Temperature Variation	Rectangular	1·00	Measured
11 Altitude	Normal	1·00	Linear
12 Vibration	Normal	0·10	Linear
13 Repetitive Transients	Normal	1·00	Linear
14 Conducted A-F Susceptibility	Normal	0·0025 (D.C. Power) 0·0055 (A.C. Power)	Linear
15 Magnetic A-F Susceptibility	Normal	1·00	Linear
16 R-F Susceptibility	Normal	0·0001 (Radiated) 0·0033 (Conducted)	Linear
17 Humidity	Normal	1·00	Linear

*Extracted from RTCA Document No. DO-131.

Reference 2 also specifies the range over which each variable must be considered to vary, and the method by which the results of testing for each variable separately are combined to give the overall centring error. In this case the apportioned standard deviation of 4·6 ft is equivalent to a centring error of 2·0 microamps. In demonstrating compliance with this requirement, some alleviation is permitted where a variable cannot have values over the full range normally assumed. In that case, testing may be limited to a more restricted range which is representative of the conditions under which the receiver will be operating in the final stages of a low-visibility approach in conjunction with ground signals of the specified quality.

Auto-pilot Accuracy – Piloting

The accuracy with which the auto-pilot can fly to the ILS centre-line (as portrayed by the ILS receiver) may depend on a number of variables, ranging from datum and gain variations in the sensors and computation through to external disturbances acting on the system. In fact, experience has shown that the main factor arises from wind effects (gusts and shears) during the final stages of the landing. In the pitch axis (flare-out) there may also be a significant effect of centre of gravity, flap position, aircraft weight, etc. The main objectives of the performance evaluation are, firstly, to show that risk overall is acceptable, and, secondly, to determine whether the effect of any variable is sufficiently critical that some limitation should be placed on the use of the system, e.g. where exceeding some value of the critical variable would result in an unacceptable risk on that occasion.

Since flight testing is expensive, it has to be limited to rather small samples, it may only include one or two particular aircraft and sets of equipment, and it may be difficult to show that it has adequately explored the effect of all variables over their full ranges. Therefore, considerable use is made of simulation to establish the performance distribution, and flight testing tends to be regarded as a means of verifying the simulation. But, of course, even though it may be possible to simulate the equipment and the aircraft with a fair degree of accuracy, the results will depend on whether or not the input models, representing the probability of encountering gusts of various shapes and magnitudes are really representative. Up until now certification procedures have used methods where wind is treated as a continuous 'noise' input. The intensity of turbulence in the simulation has been directly correlated to the 'reported' wind speed. However, over the years it has become fairly evident that this method can over-estimate the normal performance variability of the system and the increase of risk with wind speed. The limitations being placed on the use of the system, such as maximum wind speed for automatic landing, may, therefore, be too stringent. There is also the likelihood that the worst gusts for the system may in fact not be related to high wind speeds. So the wind model being used for performance assessment is being re-examined and in particular there is a need to determine whether discrete gusts, gust sequences etc., which the system must cope with safely should be specified. These would still be based on probability considerations, but might be better able to represent the more extreme, but unlikely, events.

124

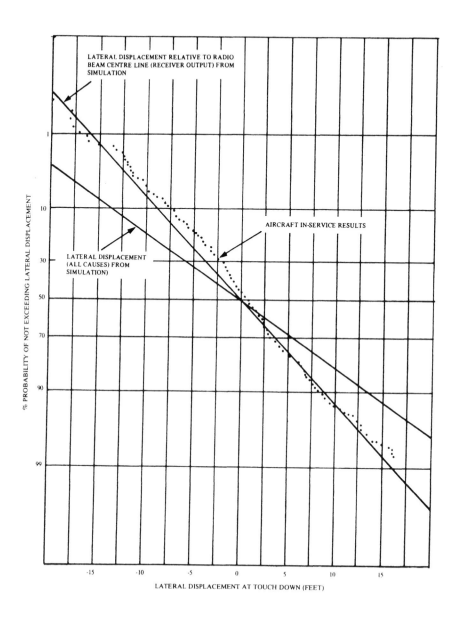

Fig. 8-2   **PROBABILITY DISTRIBUTION OF LATERAL DISPLACEMENT
AT TOUCH DOWN: COMPARISON BETWEEN SIMULATION
AND IN-SERVICE RESULTS
(611 LANDINGS WITH ONE AIRCRAFT TYPE)**

125

However, before permitting an automatic-landing system to be used in very low visibilities, the CAA requires a period of use and data collection in service to support the conclusions of the safety assessment. Fig. 8-2 shows the information relating to lateral position at touch down for one aircraft type. The dots show the in-service experience, and the unbroken lines show the simulator results relative both to the runway centre-line, and relative to the radio beam centre-line (receiver output). There are two main points to note in this figure:–

a. The close agreement between the in-service results and the simulation.

b. The achieved performance of this system is considerably better than the minimum acceptable. The standard deviation is approximately 6·5 ft as opposed to 11·2 ft allocated to it in Table 8-1.

Failures

So far, the discussion has concentrated on verifying the performance of a failure-free system and establishing any limitations necessary for its safe use. But there may be another reason for looking at performance, and that is to determine the effects of failures. Where, for example, a failure can result in degraded performance without giving a definite failure indication to the pilot, the system would continue in operation until the next maintenance check at which the fault should be found. It may, therefore, be important to determine the performance of the system in this condition to show that the risk level is appropriate to both the probability of the fault and the continued use of the system in its faulty condition.

This description has used an automatic-landing system as a typical example. The method is applicable to other systems, particularly those used for maintaining the aircraft on a particular flight path.

STRUCTURAL ASPECTS OF FLIGHT CONTROL SYSTEMS

Introduction

In this Chapter, "flight control systems" is taken to include not only the primary flight controls but also secondary controls e.g. those for flaps, slats, spoilers, trimmers, etc. Such systems often include electrical and/or hydraulic components and always include mechanical parts such as cables, rods, levers and support brackets. The purpose of this section is to discuss the integrity of the mechanical parts.

In principle, the objectives of design and the arrangement of the mechanical components in the system are just the same as for electrical and hydraulic components. In all cases the purpose is to reduce the failure probability to acceptable levels. The same approaches, such as the avoidance of common-mode and cascade failures, and the provision of redundancy, are followed. The main distinction is that mechanical parts are subject to stresses from applied external loads. Hence the title, "Structural Aspects".

The difference in the approach to airworthiness of the mechanical parts is largely due to the fact that they are stressed, and to some extent due to historical development. As regards the latter, electrical and hydraulic components have always been regarded as liable to fail, so safety assessment has been much concerned with the failure probability of redundant systems. In contrast, historically, structures were designed not to fail, as witness the popularity at one time of single spar wings. Thus, to a greater extent the design of stressed components has been concerned with establishing loading conditions unlikely to be exceeded, and with factors of safety. It is, of course, true that experience has shown that no structure is immune to failure, and increasing use of redundancy in the form of multiple load paths is now commonly practised. Thus the structural and systems outlooks have converged. But as will be seen from what follows, prediction of failure probabilities of structural and mechanical elements is not an exact business.

The Main Requirements

The main requirements, JAR-25, FAR-25 and BCAR, are similar. JAR-25, for instance, specifies that:–

> The aeroplane must be shown . . . . to be capable of continued safe flight . . . after any of the following failures.
> 1. Any single failure not shown to be extremely improbable, excluding jamming (for example, disconnection or failure of mechanical elements, or structural failure of hydraulic components, such as . . . .).
> 2. Any combination of failures not shown to be extremely improbable . . . .

These basic requirements lead to consideration of both static strength and fatigue resistance.

Static Strength

Taking as an example the primary flight controls, BCAR specify that the control system should have proof and ultimate factors not less than 1·0 and 1·5 respectively, under loads arising from:–

a. The maximum pilot effort loads. These apply to the parts of the system lying between the control column or rudder pedals and the power actuator, or through the whole system in the case of a fully manual control.

b. The loads from the power actuator; these being the maximum loads which the power actuator can produce under practicable operating conditions.

Where a system provides manual reversion in the case of power failure, the maximum pilot effort loads would also be applied throughout the system.

The requirements specify the maximum pilot efforts to be assumed based on judgement of pilots' strength. The loads from the power operation are those predicted for the particular system. The ultimate factor of 1·5 is an arbitrary figure based on experience, intended to cover uncertainties including variability of strength, and loss of strength due to cracks or corrosion or workshop errors.

It will be seen that this basis for establishing static strength does not lend itself to determining the probability of failure. However, it is true to say that experience justifies the arbitrary figures in the sense that static failures very rarely occur.

For vital systems, such as the primary flight controls, mechanical components are usually duplicated, or a standby is provided. Systems so designed are regarded as meeting the basic demand that failures be extremely improbable.

In more detail, the static strength requirements have also to be met **after** a first failure. The parts of the system which then continue to operate may be more heavily loaded than they are when the system is intact. Jamming or partial seizure of a system can also result in the application of excessive manual loads. The magnitude of the loads to be applied after a first failure depends to an extent on whether the existence of the fault is self-evident. If it can remain as a dormant fault until revealed by inspection the full loads would be expected to apply. If the failure is obvious, then the shorter period of exposure may allow some reduction of loads.

The requirements also cater for pilot error by requiring the pilot effort loads to be applied by both pilots acting together, and in opposition to each other. Jamming of controls is by no means an uncommon fault. Controls should be tested as assembled to ensure that when under load their operation is not affected by excessive friction or excessive deflections, or otherwise show a liability to jamming.

Fatigue

Liability to fatigue is a matter of greater concern than that of static failure, and is in itself justification for duplication of vital systems. If duplication is not practicable in a local area of design, then that part of the system should be so lowly stressed as to keep it clear of stress levels capable of inducing fatigue damage.

The fatigue life of a component is dependent on:–

a. The spectrum of applied loads.
b. The internal stresses resulting from the applied loads.
c. The S-N (stress versus cycles) curve for the material.
d. The scatter of fatigue life about the mean.

The S-N curves are available for most commonly used materials. The internal stresses are amenable to calculation, though tests under cyclic loading may be necessary to support calculation. The matter of greatest doubt is the loading spectrum. While generalised data based on massive flying experience are useful in the design of the main structure of an aircraft, the nature of the loading on the control system makes the spectrum more peculiar to the particular type. The loads may also vary to a greater extent from pilot to pilot. The only general guidance that can be given is that in reading across from service data the load spectrum assumed should be a conservative one. For airframe structures, a factor of 1·5 is applied to allow for a particular aircraft experiencing more frequent loads than the average. For control systems, a higher factor is advisable.

As regards the scatter of fatigue life arising from variability of material properties and dimensions, this is such that the 'safe-life' is considerably less than the mean life. Sufficient data exist about the width of the scatter band to permit the use of generalised factors to convert from mean life to safe-life. These factors, which depend on the number of tests made to establish the mean, are quoted in requirements. As an example, with three test results, the safe life is taken as one-third of the mean. This factor, together with the above mentioned allowance for variation of the loading spectrum, means a total factor of 4·5 or more on life. This may seem generous, but apparently minor deviations in manufacture can erode it significantly.

A big difference between the main airframe structure and control systems is that the former is a 'fixed' structure, while the latter is an assembly of moving parts. Hence controls suffer the additional effects of wear and tear and friction between parts.

It is largely due to these doubts as to whether a safe-life is truly safe that the need for duplication or other forms of redundancy has come about. Given an alternative path to carry the loads after a first failure, the estimation of fatigue lives is not so crucially important. However, the fact that the control system has been designed to be fail-safe does not mean that consideration of fatigue failure can be dismissed.

First, it must be remembered that if the two channels of a duplicated system are nominally identical and share equally the loads, then if one suffers a fatigue failure the second may well be close to doing so. Second, after a first failure the operative channel may be more heavily loaded, and a small increase of stress has a profound effect on fatigue life. Thus the chance of a cascade failure is much increased.

Safety achieved by fail-safe design is entirely dependent on discovering failures before they reach danger point. Before fatigue fracture occurs there is a period of time in which the crack is growing from a minute size to the point of collapse. With parts made from sheet materials, there is a prospect of detecting such cracks by inspection before danger point, particularly if the rate of growth is slow. But with 'solid' pieces, such as end fittings on tubes, the cracks may be much more difficult to see before they become of critical length. Too much reliance on inspection during routine maintenance is unwise. Only when the location of cracks is known, and specific inspections made, sometimes using special techniques, is detection by inspection to be relied on.

The upshot of this is that where the system is designed to be fail-safe by duplication, it is still necessary to consider fatigue failures with care. If failure of the dual load paths cannot be ruled out, and if the detection of a first failure before a second one occurs is in doubt, then it may well be necessary to apply safe-lives to the components.

By so doing, there is a good prospect that vulnerable parts will have been replaced before the onset of fatigue. In the event that the occasional fatigue cracks will still occur, the test programmes undertaken to establish the fatigue lives will provide some indication of the location of the likely cracks and their propagation rate, which information will assist in the setting up of inspection procedures.

Again, it is evident that the particular problems of control systems make it difficult to make any accurate numerical assessment of probability of total failure. However, if the approach to design for fatigue resistance is a thorough and conservative one, then one can expect the risk of total failure to be extremely improbable.

ACTIVE CONTROL SYSTEMS

'Active controls' has come to mean controls which have the capability of sensing external inputs and translating the messages received into appropriate control movements. Thus the sensors measure aircraft motion (pitch, roll or yaw rates, and acceleration), control surface angles, and angle of incidence to the air flow. These are analysed by the system, and the output signal then demands the control movements necessary to achieve whatever result is desired.

An existing minor application of active control is the yaw damper, which senses unwanted yawing motion and automatically applies rudder movement to damp the oscillation. In future, transport aircraft are likely to use active controls to fulfil other purposes, the two most likely developments being load alleviation and so-called 'relaxed' stability.

For load alleviation, an increase of incidence, due to a manoeuvre or up-gust, is sensed and the signal is translated into a command for upward movement of the ailerons. The sensing device could be vanes, but current research may lead to using acceleration measurement. The effect of the symmetrical movement of the ailerons is to reduce the upward air load at the wing tips and to increase it inboard, thereby reducing the bending loads on the wing structure.

For relaxed stability, in lieu of large fixed tail surfaces to provide stability in pitch and yaw, appropriate automatic movement of the elevators and rudder provides the necessary restoring action, permitting smaller and lighter surfaces.

Applied in an extreme form, active controls would permit very large reductions in strength and natural stability. While this would be attractive from the viewpoint of saving weight and drag, it would mean complete reliance on the integrity of the system. At the present state of the art, this is not regarded as feasible. Present intentions are to provide active controls of limited authority, such that after failure of the system the aircraft would remain flyable.

With active controls of limited authority, there are still considerable potential risks to contend with. First, it is necessary that, after system failures, the aircraft should have adequate strength and stability to allow a continued safe flight and landing to be made. Second, the risk of overstressing as a result of active control failure must be avoided. Third, the design of the system must produce the correct response in a wide variety of circumstances. It is evident that the safe design of active controls involves not only the normal systems approach to the electric/hydraulic/mechanical components, but also consideration of the aerodynamic and structural implications.

The provisional requirements proposed by the CAA are intended to achieve a simply expressed objective; that the probability of structural failure of an aircraft with active controls should not be greater than that of an aircraft with conventional controls. It is the practical implementation of this objective which gives rise to new problems. This can be seen by considering the case of wing load alleviation.

Conventionally, the bending strength of the wing to withstand flight loads is largely determined by applying, separately, two loading conditions. The ultimate strength is required to be not less than 1·5 times the higher of the limit loads arising from:–

a. Specified pitching manoeuvres.
b. Specified up- and down-gusts.

Experience has shown that this has resulted in adequate strength when applied to aircraft having normal characteristics, e.g. size, stiffness, aerodynamic response.

At first sight it might appear that aircraft with active controls designed to these same strength requirements would automatically be satisfactory. This may prove to be the case, but there are reasons for caution. It must be remembered that in reality aircraft are subject to combinations of manoeuvre and gust loads, and little is known about these combinations. Further, the gust model used to represent the atmosphere is a simplification derived from load measurements made on conventional aircraft. It is, therefore, uncertain whether this model will be applicable when used to determine the strength needed in wings of aircraft having unconventional responses to gusts.

An added difficulty is that the probabilities of occurrence of limit and ultimate loads are not known with any precision, so that the combining of system failure probabilities with external loading conditions is not clear cut.

While recognising the nature of these problems, the CAA has stated a number of principles, which may be summarised as follows:–

a. The probability of exceeding the limit loads should be no greater on an aircraft with active control than on a comparable one without. This will entail taking account of the probability of system failures; some failures, such as runaway, could increase the loads at the time of failure; others could leave the aircraft, after failure, with reduced capacity to withstand manoeuvre and gust loads.

b. The probability of structural failure, i.e. of exceeding ultimate strength, should be no greater on an aircraft with active controls than on one without.

c. Provisionally, with any system failure not shown to be extremely improbable, the limit loads should not exceed the ultimate strength.

In applying these principles, consideration would be given to gust models which are more realistic representations of the atmosphere than the models conventionally used. Consideration would also be necessary to ensure that combinations of manoeuvre and gusts loads do not erode the margins of strength normally present.

The use of active controls should reduce the level of repetitive loads due to turbulence, and hence be of benefit to fatigue life. However, it is not yet clear whether the fact that an active control is active could result in a different pattern of loads which might be more damaging from a fatigue viewpoint. Consideration of the fatigue effects will evidently be important not only in the aircraft structure but also throughout the mechanical elements of the system.

Plainly, there is much still to be learned about the techniques for establishing confidence in the integrity of the active control aircraft. It may well be that in the first designs to enter service statistical monitoring of the loads encountered will be a sensible precaution. This could assist in confirming, or otherwise, the assumptions made in the design stage.

DIGITAL SYSTEMS

Introduction

Up to the present, there has been relatively little experience of the application of digital techniques to safety-critical systems on civil transport aircraft. However, the next generation of large transport aircraft will use digital computation extensively in such systems as auto-pilots used for automatic landing in low visibility conditions, engine controls, active control systems, lift and drag controls, flight instrumentation, general flight management, etc., etc. After this the application is likely to be extended to various types of control configuration, where the digital computation is integrated into the main flying control system for the whole of the flight, and to structural load relief systems. It seems certain that the number of safety-critical systems using digital computation will increase rapidly over the next decade.

While, at present, there are no airworthiness requirements written specifically for digital systems it appears likely that the general system requirements described in Chapter 4 will be applied to them. However, because of the basic differences between digital and analogue computation systems, the methods of demonstrating compliance and the detail of the safety assessment may be different in some aspects. It is not intended that this Chapter should anticipate detailed airworthiness requirements or methods of demonstrating compliance, but at this stage it is useful to highlight some of the problems which are likely to arise.

It is possible to limit the failure analysis of an analogue computation system to consideration of a relatively small number of failure modes of individual hardware components. The effect on the system of a component failure can be determined precisely and its probability reasonably well estimated from experience with comparable equipment. However, the analysis does have to consider failures occurring in combination, the number of such failures considered being determined by the probability level which it is necessary to establish.

For a digital system the picture is somewhat different. Since many functions are routed through the same hardware, the effect of an individual component failure is likely to be widespread through the computer, and it may be complicated and variable depending to some extent on the nature of the computation being made at the time. Moreover, because of the discontinuous nature of digital computation, a correct outcome cannot necessarily be inferred or forecast from the correct completion of a similar computation. Thus 'bottom up' (hardware) analysis is unlikely to be of use, except for ascertaining the probability of a precisely defined narrow range of effects for specific parts of the system. 'Top-down' analysis, using a fault tree analytical approach, may be possible for most circuitry functions. However, the central processing unit is not amenable to this type of approach, and other methods of establishing integrity need to be considered.

However, with a digital system it is possible to incorporate failure-detection and monitoring circuits and devices to a greater degree, and probably with more certainty of detection than with an analogue system.

A major problem is likely to concern the 'software'. Although it cannot "fail" in the normal sense, it can cause functional failures because of errors (e.g. in logic) which are not anticipated or allowed for in the design. Unless these are revealed in analysis or testing, they will remain concealed in the system, and may cause unexpected problems in service. Also, modifications to software will almost certainly take place in service, and considerable care must be taken to ensure that no unwanted side effects are introduced in this process.

Because of the very small dimensions of the components and the consequent high density of packaging, there is a potential for complex failure modes of a largely unpredictable nature, so that electrical and mechanical segregation, in relation to functions, is of great importance in the basic design.

As with analogue systems, the overall safety will be dependent, to a large extent, on the structure or 'architecture' of the system and the way it is installed relative to the threats from other parts of the aircraft. The Preliminary Hazard Analysis should study the system architecture and segregation within the installation. Aeronautical Radio Inc. (ARINC) Paper 425, "Digital Flight Guidance and Control System Architecture" (Ref. 4) is a relevant document.

The safety assessment may be broken down on the lines described in the previous Chapter as follows:–
a. Definition of System.
b. Preliminary Hazard Analysis.
c. Definition of Safety Objectives.
d. Software verification and validation.
e. Failure analysis of:–
    i. Hardware.
    ii. Firmware (built-in processor instructions).

Software

The term 'software' is used to represent the sequence of instructions which are stored in the memory of a computer and are executed by the processor to provide a programme associated with data movement between memories, processors and input/output devices.

The main difficulty lies in verifying the correctness of the software and validating that it does, in fact, cover all of the intended conditions needed for safe functioning of the system, bearing in mind that the contents of the software will probably be complex and recorded in a special language.

A prerequisite for ensuring that the software is free from errors and is valid in relation to the functions of the system, is to have well-ordered and disciplined control procedures which define the objectives, rules and methods which will be applied during the software development programme. In order to do this effectively it would appear necessary to break down the software, as far as is practical, into discrete modules which can be tested, analysed and understood at an engineering level. The more intricate a programme becomes the greater is the temptation to make more efficient use of the software and economise on hardware; this may lead to system safety considerations being overlooked.

The Airworthiness Authorities are likely to insist that vendors of equipment and aircraft constructors define company procedures and processes which strictly control the system design specification, software architecture, flow charts at module and system level, and checking procedures both at programme and system engineering levels, in addition to procedures for testing and analysis. The effort involved in this process is likely to be large.

It is not envisaged that it will be practical at this stage to predict the probability of a software error since this, essentially, starts as human error.

However, even with the extensive controls and detailed examination described there will remain the question – "Have we eliminated all sources of error likely to degrade the level of safety?". The credibility of the answer will depend on the degree of complexity of the software and the extent of knowledge of it and the authority of the system. For systems such as active controls with moderate authority and auto-land, it is usually possible to conclude that there are no design errors prejudicing the required level of safety of the system, mainly because of the limited areas where the error could be catastrophic and the relatively short periods of risk exposure. However, for those systems where availability is the prime criterion (e.g. 'fly-by-wire') the problems are more acute and will probably need additional provisions such as the use of independent back-up systems.

With micro-processors, or 'on-chip' processors, the problem becomes more intense because the manufacture and control of the 'chip' will probably be the responsibility of a manufacturer who is not involved in the aircraft project, and the software may not be sufficiently 'visible' for a safety assessment to be made. In such cases some form of independent monitoring is essential where the effects of failure and error may be hazardous or catastrophic.

One way of dealing with possible software design errors is by the use of dissimilar software. However this in itself gives rise to its own problems. First one has to decide where the dissimilarity begins and ends. For example, if the programme is only dissimilar in respect of encoding, errors which exist in the design specification will still be propagated through the system: so, in order to ensure that this does not happen, it is necessary for the dissimilar software to be generated and controlled independently and to ensure that it matches with the main software, and that each set of software is subjected to the intensive verification and validation processes referred to above. Fig. 8-3 shows an example of the stages used in a particular dissimilar software development process for a critical system employing an active lane and a monitor lane.

One has to be very careful, in systems for which continuous availability is essential, that the use of precautionary techniques such as those described, while lessening the danger of malfunction, do not prejudice the availability of the system. It is important to bear in mind that one of the advantages of using digital techniques is the avoidance of interlane disparities.

There are, therefore, in the design of critical digital software, numerous challenges to the ingenuity of the designer. At this stage it would be unwise to be dogmatic about the solutions to the problems.

Failure Analysis

A failure analysis will be needed on the total system to demonstrate compliance with the airworthiness requirements.

The analysis of the processor hard-wired instruction set (Firmware) including any micro-programme and control store, may be analysed for failures and effects by using a combination of simulation and theoretical analysis. There are some characteristics of processors which can assist in analysis for example:–

a. Most failures have an immediate and significant effect.

b. Many instructions are executed numerous times during one pass through the application programme.

c. Most functions and components are frequently exercised.

An assessment should be made of all of the common circuit elements. Many of the failures (e.g. to the arithmetic logic devices) will lead to a massive disruption of programme flow which becomes readily detectable. However, this alone may not be sufficient for the safety of the system, so that the analyst's task is to ensure that all such failures are detected by the monitors and appropriate corrective action taken, either automatically or by clearly defined crew procedures, as appropriate.

More difficult failure cases are the low-error-rate failures associated with, for example, an individual "bit" error within a micro-programme control store. This type of failure may affect only one type of operation performed

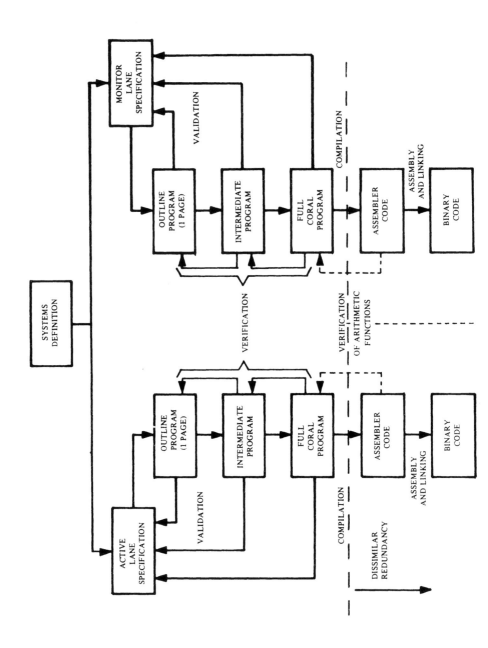

Fig. 8-3   **DISSIMILAR SOFTWARE DEVELOPMENT PROCESS**
(Courtesy of Smiths Industries)

by the processor. Such cases need a monitoring process controlled by hardware or software so that the failure can be detected before the function is needed.

Failures in memory, either Random Access (RAM) or Read Only (ROM), fall into two categories, those that involve common failures and affect all memory addresses and those that involve the failure of an individual "bit" in a memory cell or to the "bits" in groups of addresses. In the first case, the error rate is likely to be very large and detection relatively simple. In the second case, the error rate is likely to be very low and can only be detected by properly designed monitors (e.g. memory check sums).

Failures in the input/output section, analogue/digital conversion etc., are amenable to similar arguments.

An important element in the analysis, is the search for common-mode failures, which has been dealt with at length in Chapter 6, but here the problem is further complicated by the possibility of common errors in software.

The most difficult part of the analysis relates to the processor which is not amenable to traditional 'top-down' approach or traditional Failure Modes and Effects Analysis (FMEA) and, therefore, will require particular theoretical analyses to be performed which will be dependent on the software application programme and monitoring. Consideration has to be given to remote failures of devices which may only cause an effect with a specific input data combination; such faults may remain dormant for considerable periods of time and then cause an incorrect output when the critical combination occurs.

Performance Analysis

The performance analysis of digital systems should not differ substantially from those techniques used for analogue systems.

References

1 International Standards and Recommended Practices: Aeronautical Telecommunications (Annex 10 to the Convention on International Civil Aviation), ICAO, Montreal.
2 "Document No. DO-131 – Minimum Performance Standards – Airborne ILS Localizer Receiving Equipment", Radio Technical Commission for Aeronautics (RTCA) Washington D.C.
3 CAP 208 Volume 1, Part 2 – "Airborne Radio Apparatus: Minimum Performance Requirements", Civil Aviation Authority, (UK).
4 Paper 425 – Digital Flight Guidance and Control System Architecture, (Aeronautical Radio Inc. (ARINC) (USA).

9
RECORDS AND REPORTS OF SAFETY ASSESSMENTS

THE PURPOSES OF RECORDING AND REPORTING

Previous Chapters have described the various studies and analyses involved in safety assessment. Adequate written records are essential. First, they are important in the design process as a means of ensuring that the assessment has been properly made in all respects. Second, they are needed as a basis for reports which will be required by the Airworthiness Authorities for certification or approval purposes. Third, they will probably be needed later when the systems are in service and it becomes necessary to check the suitability of modifications or changes in procedure.

The records of the assessment of a complex system are unavoidably voluminous. For this reason their arrangement and form of presentation is important so as to secure not only a comprehensive, but also a comprehensible, total record. The method of presentation adopted by a particular firm will depend on the nature of the system being assessed and on the organisational arrangements within the firm. What is important is that there should be an 'in-house' code of practice adopted by all concerned with a particular project, such as should ensure full and unambiguous communication between them.

An important feature of the records is that they clearly identify critical features of the system, as such features may need particular attention in design, manufacture, and in specified crew and maintenance procedures.

SCOPE OF RECORDS

As has appeared in earlier Chapters, a safety assessment is likely to include some or all of the following aspects:–
a. Preliminary Hazard Analysis.
b. Definition of Airworthiness Objectives.
c. Zonal Analysis.
d. Studies of particular cascade and common-mode failures (e.g. engine burst studies).
e. Failure Modes and Effects Analysis (Fault/Failure Analysis).
f. Performance Analysis.
g. Reports on testing; bench, simulator and flight tests to confirm assumptions made in analysis.
h. Reliability studies based on testing, manufacturers' data or reference to previous experience.

It may often be found most convenient to make separate reports on many of the above aspects of the investigation. However, it is important that the whole should be fully cross-referenced for ease of use and to ensure that no gaps are inadvertently left in the evidence.

Generally, it will be found best to provide a master summary document to 'tie together' the whole assessment, and this will make cross-reference to the individual supporting reports.

THE MASTER SUMMARY

Again, the exact content of the master summary may vary from case to case, but its main purpose would be to provide:–

a. A definition of the system and its functions.
b. A statement of the Safety Objectives.
c. A description of the means used to show compliance with the objectives and a summary of the conclusions reached, with particular emphasis on critical safety areas in the system.
d. A summary of the limitations in service on which the assessment is based, e.g. flight-crew checks, maintenance procedures and frequency of checks.

DOCUMENTS REQUIRED BY AIRWORTHINESS AUTHORITIES

Reference should, of course, be made to particular authorities to determine the documentary evidence required for certification. At the time of writing, BCAR asks for a summary to be prepared containing the following information; JAR covers much the same ground.

a. A definition of the system and its functions.
b. A list of the equipment of which the system is comprised, and a definition of its standards at the time of certification.
c. A statement of the Airworthiness Objectives.
d. Summaries of analyses and tests.
e. Conclusions, showing that the Airworthiness Objectives are complied with, and listing any necessary conditions, which are associated with compliance, e.g.:

    i. Limitations on the use of the system.
    ii. Flight-crew procedures and checks.
    iii. Maintenance procedures and periods.
    iv. Allowance for unserviceable parts (e.g. Master Minimum Equipment List).

NOTE: The associated conditions should be cross referenced to such documents as Flight Manuals and Maintenance Schedules.

APPROACHES IN CURRENT USE

In compiling safety assessments for some Joint European projects, aircraft constructors have adopted a format dealing with the following subjects. While it is not suggested that this is the only way of doing the job, the following is based on what has been done and is offered as an illustration.

Chapter 1. General:
(a) Definition and description of system.
(b) Interfaces.
(c) Airworthiness objectives.

Chapter 2.   Conclusions, Including:

(a)   A comparison between the objectives and the results of the assessment.
(b)   A discussion of the significant conclusions.
(c)   Failure cases which require special flight-crew action, but cannot (because of the hazard) form part of flight-crew training.

Chapter 3.   Component Failures and Failures External to the System:

(a)   A list of component failure rates or modal failure rate with details of their origins.
(b)   A list of relevant failures in other systems.

Chapter 4.   Failure Conditions:

(a)   A tabulation of failure conditions, not including flight-crew errors, describing the effects of the failure condition as seen by the flight-crew, their consequent actions, the probability of the failure, and cross referenced to the individual parts of the analysis from which the failure conditions are derived.*
(b)   Human and installational failures/errors, including studies of:
   (i)   Crew actions
   (ii)   Maintenance errors
   (iii)   Installational failures (e.g. caused by environment, cascade failures).
(c)   Dependence diagrams, fault tree diagrams or tabular representations etc., of failures, or combinations of failures, for each failure condition.
(d)   Sensitivity to variations in failure rates, where these cannot be predicted with confidence.

Chapter 5.   Check Periods Assumed in the Analysis.

Chapter 6 etc.   Detailed Derivation of Sub-system Modal Failures and Their Rates.

RECORDING FAILURE MODES AND EFFECTS ANALYSIS, AND PERFORMANCE ANALYSIS

One of the more difficult records to prepare is that concerned with Failure Modes and Effects Analysis, and Performance Analysis. The records and summaries require much care to prevent important aspects becoming lost in the welter of detail. S.A.E. Document ARP 926A (Ref. 1) is helpful, and it provides examples of format of presentation.

Table 9-1 gives a particular example of a summary relating to the failure conditions of a fuel system for a four-engined turbo-jet aircraft. The failure modes producing the failure conditions include those to fuel pipes, fuel tanks, recuperators, electrically-driven and engine-driven pumps, fuel cocks, fuel jettison system, non-return valves, electrical system, etc., and will have been documented and cross-referenced in the summary. The summary includes flight phase, crew warnings and crew drills, as well as the details of the effects of the failure conditions.

It will be noted that Table 9-1 deals only with the loss of fuel supply. However, some of the contributing failures will have other effects: for example, broken pipes and leaking tanks may well produce fire risks; these will be included in other relevant failure summaries (e.g., for fire risk).

---

* See Table 9-1.

TABLE 9-1
SUMMARY OF FAILURE CONDITIONS – LOSS OF FUEL SUPPLY

Failure Condition	Flight Phase	Indication	Crew Action	Category of Consequences	Probability per Hour of Flight	Reference to Failure Modes
(1) Loss of fuel supply to any one engine	Any phase except take-off	Master Warning System (MWS) plus audible warning Fuel low pressure warning Fuel flow meter Loss of engine power	Engine shut-down procedure Possibly aborted flight	Minor	10^{-5} Remote	Quote references to items in FMEA producing failures (should include relevant electrical failures) with references to fault trees, dependence diagrams, etc.
(2) Loss of fuel supply to any one engine	Take-off, up to 100 knots	MWS no audible warning Fuel low pressure warning Fuel flow meter Loss of engine power	Emergency STOP Procedure	Minor	2×10^{-8} Extremely Remote	— ditto —
(3) Loss of fuel supply to any one engine	Take-off, 100 knots to V_1	MWS no audible warning Fuel low pressure warning Fuel flow meter Loss of engine power	Emergency STOP Procedure	Major	4×10^{-8} Extremely Remote	— ditto —
(4) Loss of fuel supply to any one engine	Take-off, V_1 to (V_1 + 20 secs)	As (1) above	Take-off followed by engine shut down procedure	Major	10^{-8} Extremely Remote	— ditto —

Failure Condition	Flight Phase	Indication	Crew Action	Category of Consequences	Probability per Hour of Flight	Reference to Failure Modes
(5) Loss of fuel supply to any one engine during negative acceleration	Any phase except take-off	MWS plus audible warning – recuperator caption – Fuel low pressure warning	Relight	Minor	10^{-8} Extremely Remote	Quote references to items in FMEA producing failures (should include relevant electrical failures) with references to fault trees, dependence diagrams, etc.
(6) Loss of fuel supply to any two engines during take-off	Take-off 0 to 100 knots 100 knots to V_1 V_1 to (V_1 + 40 secs)	As for (2) for two engines As for (3) for two engines As for (4) for two engines	Emergency STOP procedure Emergency STOP procedure Not applicable	Minor Major Hazardous – Catastrophic	$\leqslant 10^{-9}$ Extremely Improbable	— ditto —
(7) Successive loss of fuel supply to any two engines	Any phase except take-off	As (1) for two engines	Engine shut-down procedure	Major	2×10^{-10} Extremely Improbable	— ditto —
(8) Successive loss of any fuel supply to three or four engines	All phases	MWS plus audible warning Fuel low pressure warning, etc.	Close cross-feed valves. Lower emergency ram-air turbine Attempt to relight	Hazardous – Catastrophic	$\leqslant 10^{-9}$ (without crew error) Extremely Improbable	— ditto —

CONCLUDING REMARKS

As will be seen from the foregoing, there are various ways in which the designer may choose to arrange the records and reports of an assessment. To reiterate, whichever way is chosen, it should:–

a. Be comprehensive, covering all aspects of the assessment.

b. Be such that the basis for the conclusions reached can readily be traced back through the supporting documents.

c. Be comprehensible to all concerned.

d. Be capable of amendment.

Reference

1 ARP 926A "Fault/Failure Analysis Procedure", SAE Aerospace Recommended Practice, Society of Automotive Engineers (USA), May, 1979.

10
POST CERTIFICATION

GENERAL

This book is mainly concerned with the principles of safety assessment in the design and certification stages. However, when an aircraft type is in airline service and accumulates experience the safety assessment can continue to be a valuable tool.

When an aircraft enters service it may be used by many different operators in different countries, each country having its own Airworthiness Authority, so that it is difficult to see a generally agreed system of using safety assessments in service, particularly in relation to maintenance tasks, being developed quickly. In addition, most operators are using well established methods of recording information and establishing maintenance tasks and periods, and are reluctant to alter or add to these methods except in exceptional and obviously justifiable cases. Nevertheless, if the information derived from the aircraft constructor's safety assessments and provided to operators is limited to critical features and is presented in clear and usable form, there is every likelihood that the application of safety assessment methods in service will be increasingly adopted. There is an obvious need for close liaison between aircraft constructors, operators and Airworthiness Authorities in this field to produce generally acceptable and consistent methods of approach.

The aspects of operation where it is particularly necessary for the operator to make use of safety assessments are:–

a. Modification to design.
b. Changes of operating procedures.
c. Maintenance tasks and their periodicity.
d. Allowable deficiencies.

These are briefly discussed below. This Chapter also discusses monitoring reliability in service and the use of alert levels, the general benefits of adequate systems for reporting occurrences, and the response of Airworthiness Authorities and operators when hazardous situations are revealed in service.

MODIFICATIONS TO DESIGN

Where modifications are made to the design of critical aircraft systems, reference to the safety assessment is considered to be an essential continuation of the design process. The assessments should be up-dated since even if a modification does not affect the conclusions of the assessment, it may have an influence on future modifications to the system.

One of the difficulties associated with modifications is that they may be introduced by sub-contractors or operators in order to improve ease of manufacture or maintenance, but may prejudice some of the safety features of the system (e.g. segregation, introduction of dormant failures). It is, therefore, important that the aircraft constructor is aware of any such modifications so that he is able to assess and advise on their effect on safety.

CONFIGURATION CONTROL

With such systems as automatic-landing and stall-identification where there is a high level of risk and the systems are new or complicated, the CAA and other Airworthiness Authorities have applied a method known as **"configuration control"**. In this, all changes to the system and its components and any proposals for changes to related maintenance tasks are referred to the aircraft constructor for agreement before they can be implemented.

Such changes may derive from the component manufacturer, the aircraft constructor or the operator. The appropriate Airworthiness Authority will only accept such changes on the basis of confirmation from the aircraft constructor as to their acceptability. While this method has proved very effective, it has been limited to a small number of particularly critical systems. There are obviously various degrees to which such control can be applied, depending upon how critical the system is.

CHANGES OF OPERATING AND CREW PROCEDURES

Amendments to operating and crew procedures, including training, of a kind which touch on the assumptions made in the aircraft constructor's initial safety assessment, need to be subject to further safety assessment. This is particularly important if critical features are concerned. Flight Manuals and other crew procedure documents should accurately reflect the intended changes.

MAINTENANCE TASKS AND THEIR PERIODICITY

When an aircraft is in service, the maintenance tasks and their periods of application are adjusted by each operator according to its experience and to the combined experience on aircraft of the type. Some of the methods used for the control of maintenance which are relevant to systems using redundancy techniques are described in CAA document CAP 418 – "Condition Monitored Maintenance – An Explanatory Handbook" (Ref. 1).

Where the nature and periodicity of the maintenance tasks is shown by the safety assessment to be critical it is necessary for the aircraft constructor to make this clear to operators, so that when adjusting their maintenance schedules they are able to continue to work within the safety objectives for the system. This is particularly so in the case of checks to detect dormant failures (passive and undetected failures) (see Chapter 5).

Currently, on some aircraft systems there has been a form of limited maintenance control, in which a list of tasks derived from the Safety Assessment is prepared, and this is mainly concerned with 'hidden functions' involving dormant failures. This list becomes part of the appropriate maintenance documents. The identified tasks can be controlled in a manner separate from those normally used for the escalation of check periods. When changes are made to critical check periods these are referred to, and agreed by, the constructors. Such control has varied in depth according to the associated degree of hazard, and has been limited to those tasks associated with those items the failure of which could result in Hazardous and Catastrophic Events. This method is less rigorous than the "configuration control" described above.

This aspect of the application of safety assessment needs considerable negotiations between constructors and operators, and it is difficult to devise global solutions. It is necessary for constructors to provide very clear information to the operator in a digestible form and to fit this, as far as is practical, with the established methods of control.

One way of reducing the problem is to reduce the number of dormant failures. This can be done, particularly in the case of avionic systems, by built-in test equipment, or by devising systems where dormant failures are eliminated as far as is practical.

With mechanical systems these objectives are difficult to achieve but are worthy of investigation.

ALLOWABLE DEFICIENCIES

With modern aircraft in which redundancy techniques have been employed, it is usual to provide 'extra-redundancy' in some systems to enable the aircraft to take-off and complete a flight with adequate margins of safety even if, say, one channel of a system has failed during a previous flight. This procedure avoids the aircraft being delayed away from its main base where repairs can more readily be made. Minor deficiencies, even without the provision of extra-redundancy, which do not too seriously affect safety, may be acceptable for the occasional flight. In both instances, safety assessment may well be a useful guide to what is acceptable.

When an aircraft goes into service the aircraft constructor in co-operation with operators prepares a Master Minimum Equipment List (MMEL) which defines the minimum standard for take-off, and the limitations (e.g. weather, length of flight, speed, altitude) associated with these minimum standards. Each operator then produces his own Minimum Equipment List (MEL) appropriate to his own routes and procedures within the limitations defined by the MMEL.

These documents will be mainly concerned with allowable deficiencies in systems, and depending on in-service experience the aircraft constructor or operator may wish to amend them.

Where items are included in the MMEL or MEL, account should be taken of them in the safety assessment. An example of a calculation involving an allowable deficiency is given in Appendix 5-2.

MONITORING RELIABILITY IN SERVICE

It is common practice to analyse records of failures of components during in-service operation with a view to detecting changes of rates of occurrence.

For instance, the number of failures and the number of component-hours might be counted on each of twelve consecutive months. The failure rate for each month could then be calculated, and this would provide a fair indication of the mean failure rate. The failure rate experienced in the next, and subsequent months, could then be compared with the previous mean rate.

The problem is that the twelve individual results collected will display some scatter about the mean. In comparing a subsequent result it is, therefore, necessary to decide whether its departure from the mean is likely to be no more than ordinary scatter or an indication that a real change has occurred.

For this purpose, it is usual to establish an Alert Level above which, in the case of deterioration, action would be justified to try to discover the cause of the change.

To illustrate by example (see Fig. 10-1(a)) suppose that the rate per 1,000 hours for each of twelve periods is as follows:–

$$1.2, 0.9, 1.1, 0.7, 1.0, 1.2, 0.8, 1.4, 1.1, 0.9, 0.7, 1.0.$$

The mean rate for these results is 1.0. The degree of scatter can be calculated by finding the standard deviation, and in this example the s.d. $= 0.22$.

If we make the assumption that the scatter is in accordance with the Normal Distribution, it follows that on 1 occasion in 100 the rate would be expected to be in excess of

$$\text{Mean} + 2.3 \text{ s.d.}$$
$$1.0 \quad + 0.5 = 1.5$$

If, therefore, we fixed the Alert Level at a rate of 1.5 per 1,000 hours, it would mean that a result in excess of this level would be highly unlikely to be pure chance. Put the other way, we would be 99% confident in concluding that a departure of this magnitude indicated that real deterioration had occurred.

Clearly the Alert Level can be set at any level desired. If it is set high, relative to the intrinsic scatter, then it will be a reliable indicator that trouble is real. If marked deterioration is of no great consequence to safety, then a high Alert Level is preferable, as it avoids too many false alerts. On the other hand, if it is important to have early warning of deterioration, then the Alert Level should be set low, accepting that it will result in more frequent false alerts.

The above method relates the Alert Level to the scatter of results actually experienced. It is possible that, in a component the failure rate of which is of critical importance, the design assessment will have shown a 'never-exceed' rate. In this case an Alert Level would be set to ensure warning of approach to this design figure.

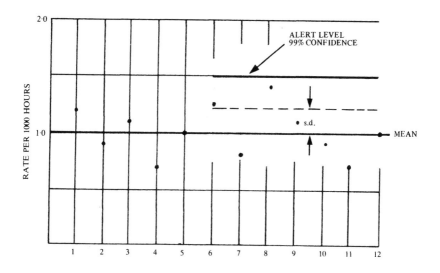

Fig. 10-1(a)   EXAMPLE 1: ALERT LEVEL BASED ON
KNOWN SCATTER

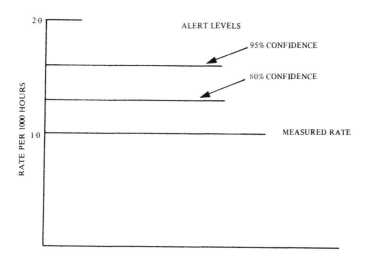

Fig. 10-1(b)   EXAMPLE 2: ALERT LEVELS BASED ON
POISSON DISTRIBUTION (10 FAILURES "EXPECTED"
IN PERIOD)

There is also an alternative approach to the matter of establishing Alert Levels, applicable when there is inadequate service data to judge the natural scatter. This method is based on the Poisson Distribution described in Chapter 5. The Poisson Distribution has built into it an assumption about the degree and shape of the scatter curve related solely to the mean value.

Again to illustrate by example (see also Fig. 10-1(b)) suppose that in 50,000 hours a total of 50 failures has been counted giving a mean failure rate of 1 in 1,000 hours. In a subsequent period of 10,000 hours the 'expected' number of failures would be 10. Assuming the Poisson Distribution to apply, reference to Poisson curves shows that when the expected number is 10, there is a probability of:–

a. 0·2, that 13 or more failures will actually occur (1·3 per 1,000 hours).

b. 0·05, that 16 or more failures will actually occur (1·6 per 1,000 hours).

Thus, if the Alert Level is set at 13 failures we can be 80% confident that the appearance of more than this number is an indication of deterioration. Similarly, a level of 16 gives a confidence of 95%.

As illustrated earlier, when the expected number is small, say less than 10, the Poisson Distribution is markedly unsymmetrical. Its use thus tends to set Alert Levels very high relative to the mean. In such circumstances there is much more doubt about the validity of the use of this approach.

In the longer term, as more data accumulates, a trend may be discernible, either of improvement or deterioration. If this is a genuine trend, and it is always helpful if the cause can be identified, action to remedy deterioration may well be desirable. In any case, given more information, the mean and Alert Levels can be re-established using one or other of the above methods.

REPORTING OCCURRENCES

Most Airworthiness Authorities make arrangements whereby 'Occurrences' in air line service are reported to a central Organisation, and then disseminated to others who need to know. Occurrences include defects and failures as well as incidents attributable to human error.

In some countries reporting is voluntary and in others it is obligatory. In addition, arrangements are often made for the exchange of information between operators, and between aircraft constructors and operators. The primary purpose of such activities is to assist in the day-to-day problems of ensuring safety, and as such they are very valuable.

Information derived from reports of incidents and defects can, in addition, be of value in design and safety assessment. For instance, provided it is properly analysed, it can give data on frequency of occurrence and on features of design or operation which prove unsatisfactory.

For lessons fully to be learned from Occurrences, it is clearly necessary that:–

a. all, or at least the great majority of, Occurrences be reported;

b. action be taken to establish the probable cause(s) of Occurrences;

c. consideration be given to the possible consequences of the repetition of the Occurrence in a variety of circumstances other than those which happened to apply in the particular reported case.

As regards reporting, it is generally found that reports on aircraft defects are readily made available, under both mandatory and voluntary systems. There is more difficulty in ensuring a free flow of reports on human error incidents. This is no doubt due to the understandable reluctance of the individual to admit to error, perhaps coupled with anxiety that admission will result in some penalty. In rare cases, a crew error can be so grave as to cast doubt on the fitness of the person making the error, but in the vast majority of cases this is not so. It is clearly difficult for a Licensing Authority to give a general amnesty to free everyone who reports from the risk that some penalising action may be taken. However, the value of reporting is so great that it is highly desirable that both Authorities and the management of operators should refrain from imposing penalties in all but the most exceptional circumstances. A climate should be created which encourages free reporting.

The USA has introduced a novel system of inviting reports to be made to an independent third party, NASA, with a guarantee that the reporters name will not be revealed. This seems to have increased the flow of incident reports. However, in this system, enquiry into the validity of the report is precluded, so it is difficult to take immediate remedial action. Nevertheless, such a system will indicate general trends.

An initial report serves little more than an alert that something has happened. To benefit from knowledge of the Occurrence necessitates considerable effort to establish the probable cause. With mechanical trouble, which on the surface seems of minor consequence, the operator may do no more than replace the failed component. Removal rates are often a poor guide to genuine failure rates. With crew error incidents, there is the problem of obtaining an accurate account of the event, as the individual, even if willing to admit to the error, may have a hazy recollection. Incident investigation is a matter which deserves, but does not always receive, specialist treatment.

Finally, when probable causes have been established, there remains the need to study the possible consequences of the same event recurring, but in other surrounding circumstances. From such studies, the risk of an apparently harmless incident becoming an accident may be foreseen and thus avoided. Again, this is a specialist activity benefiting from bringing to bear on the problem different talents, for instance, pilots, engineers, and aero-medical persons.

As will be seen, adequate follow-up of Occurrences requires significant effort with consequential costs. Nevertheless, the potential benefits justify effort, perhaps more than is customarily given. A fully understood incident, if acted on, can clearly avoid a future accident. Though the difficulties of undertaking such investigation have been mentioned above, it should be remembered that after an incident the chief witnesses survive to describe the event whereas after a fatal accident the prospects of identifying the cause of human error are slender.

In the tasks of design and the making of safety assessments, allowance for human error is one of the more uncertain areas. Study of the human error content of incidents can help in various ways. The simple fact that the human being has behaved in a particular way is instructive. Knowledge that particular types of error are frequent or infrequent is of help. Some idea of the reaction times in responding to warnings, or other inputs of information, may be gained. Knowledge of human behaviour, derived from actual experience in service, provides designers with clues to enable designs to be improved so as to render errors less likely. As has been said earlier in this book, as many, or more, accidents occur from the improper use of a properly functioning system than from system failure itself. This justifies making the best possible use of the information which is there to be had.

ACTION FOLLOWING HAZARDOUS SITUATIONS OCCURRING IN OPERATION

A not infrequent problem for Airworthiness Authorities and operators, concerns the actions to be taken following experience of troubles in service. These troubles may be one or more accidents, or a repetition of failures of particular components. Usually, such events lead to immediate practical actions, for instance, inspections on other aircraft in the fleet. But in the early stages more serious action such as modification may be in doubt. Time is needed to investigate thoroughly, and after this further time may be needed to manufacture and instal modified components. Thus, following the experience of trouble, decisions are necessary whether or not to continue flying, and, if flying does continue, for how long unmodified aircraft can properly be allowed to fly. These decisions depend on a judgment of the magnitude of the risk. When the existence of an above-normal risk is known, there is a clear obligation to decide whether it is acceptable. Such decisions may be influenced by external considerations of a legal or political kind, but this is not a matter for this book. In what follows, it is the business of assessing and judging the 'technical' acceptability of risk which is discussed.

As was mentioned in Chapter 3, when above-normal risks are known to exist, two main considerations arise. First, from the viewpoint of protecting the occupants (passengers and crew) whether the probability of accident is such that even a single flight should not be allowed. Second, if flying does continue for a period, whether the cumulative added risk is such as to prejudice the long-term accident rate over the fleet life of the aircraft.

It may assist to focus ideas to invent an example. Suppose a new type of aircraft has accumulated 1·5 million hours when a serious incident occurs. The indications might be that the cause of the incident was inadequately controlled maintenance, and flying would be permitted to continue with more closely defined maintenance procedures. Suppose that after a further 0·5 million hours a second similar failure occurs, from which it is concluded that the design itself is at fault. Suppose, further, that the failures appear to be random and not associated with age of the component.

Then, for this example, 2 failures in 2 million hours suggest a rate of 1×10^{-6} per hour. Consideration of the hazard arising from the failure might suggest that 1 case in 10 could turn out to be fatal; thus the fatal rate appears to be 1×10^{-7} per hour. Now with only 2 failures, we cannot be at all sure of this deduction. In Chapter 5, referring to small samples of evidence, it was shown that by assuming the Poisson Distribution, we could have 95% confidence that if 2 failures occurred in a given period, then not more than 5 failures would occur in other similar periods of exposure. Thus, to err on the safe side, we could assume that the fatal accident probability might be as high as 2.5×10^{-7} per hour.

Such a risk would be much higher than the target assumed in the design stage, so the need for an improving modification would be clear. The more difficult decisions would be whether, pending modification, the aircraft should continue to operate, and if so for how long.

As regards the 'grounding decision', though the risk in itself appears high, it should be viewed relative to the total risks which are, in any case, normally present. A normal fatal accident probability from all causes of accident can be taken as of the order 10^{-6} per hour. Thus the **added** risk, even on pessimistic assumptions, is no more than 25% of the risk which is normally present. Remembering the variation from one type to another above or below the notional 10^{-6} level, the added risk is relatively small.

Consider next the cumulative risk. An individual aircraft might typically have a life of 30,000 hours, probably more in future. Suppose that the time to embody modifications was set at 500 hours, something like two months flying, then the cumulative added risk would be $2.5 \times 10^{-7} \times 500$. This can be compared with the whole life normal risk of $1 \times 10^{-6} \times 30,000$. The added risk is about 0.2%, and would be negligible relative to the whole life normal risk.

The CAA has given some thought to the idea of establishing criteria based on considerations illustrated in the above example. The proposals, dated September 1978, are given in a paper written for discussion with, and consideration by, outside bodies. The CAA paper emphasises the point that the arithmetical analysis of the risks should not delay the immediate practical actions which follow the appearance of trouble. When, however, such steps have been taken and it appears that some added risk remains, the method gives some guidance on how long such additional risks could be tolerated.

The CAA Paper suggests two criteria. The first, is that the aircraft be grounded, except perhaps for a flight back to base, if the catastrophe risk arising from a particular fault exceeds 30×10^{-7} per hour. The second, sets the maximum cumulative risk from the particular fault at 1×10^{-4}. Thus, the time for which the added risk would be permitted to be present would be inversely proportional to the level of risk. This is shown in Table 10-1.

It will be seen that if, in the whole life of the type, there were as many as ten emergency periods, and if in each case the time period permitted for modification action were determined by the above mentioned limits, then the total added cumulative risk would be 10×10^{-4}. This can be compared with the 'normal' whole-life risk of say 1×10^{-6} per hour for a life of 30,000 hours, i.e. 300×10^{-4}. Thus the added risk is no more than 3·3% of the normal risk. It has to be remembered that there could be corresponding emergency periods, following operational incidents, when an added risk was tolerated during the period of time that flight drills or training were being revised. However, even allowing for this extra factor, it seems that the cumulative added risk during emergencies is not likely to prejudice seriously the long term rate. The actual periods which emerge, varying from one year for minor troubles, to a few days for serious risks, line up with past practice, and can be regarded as a quantification of what has, hitherto, been a qualitative judgment.

TABLE 10-1
RELATIONSHIP BETWEEN TIME AND LEVEL OF RISK

Catastrophe Probability of Added Risk (per hour)	Permitted Period (hours)	Approximate Calendar Time (based on 3,000 hours per annum)
$0\cdot3 \times 10^{-7}$	3300	1 year
1×10^{-7}	1000	3 to 4 months
3×10^{-7}	330	6 weeks
10×10^{-7}	100	10 days
30×10^{-7}	33	3 days*

*Note that the first criterion prohibits flying at this risk level, except for the possibility of return to base.

As regards the maximum permitted risk of 30×10^{-7} per hour, this represents an added risk of as much as 3 times the normally present risk on aircraft with a good safety performance (i.e. with fatal accident rates of around 1×10^{-6} per hour). However, some currently certificated transport aircraft, generally the older types, have fatal rates around 3 to 5×10^{-6} per hour. It would be technically difficult to justify a more restrictive criterion which grounded defective aircraft which were operating at risk levels no higher than is accepted as the norm for other existing aircraft. As time goes on, and the general level of risk reduces further, it would then be reasonable and desirable for the criterion for grounding to be correspondingly reduced.

Reference

1 CAP 418 – "Condition Monitored Maintenance – An Explanatory Handbook", Civil Aviation Authority (UK).

11
CONCLUDING OBSERVATIONS

ORGANISATION

General

The application of system safety assessment methods should not be regarded as just a method of demonstrating compliance with airworthiness requirements, but should be a continuous process applied throughout the design stages. If the assessment is delayed until a late stage in the design, this will almost certainly lead to belated and expensive modifications which are often only compromises. Although not required by current airworthiness requirements, the use of Preliminary Hazard Analyses leading to a clear statement of the particular airworthiness objectives for each system at an early stage of the design is an essential step in the process.

In order to ensure a consistent and well co-ordinated approach within the design organisation, and at the same time to ensure that major design aspects are subject to cross-checking independently of the particular system designers, it is necessary to have a form of continuous surveillance exercised from outside the mainstream of the design process. This could well take the form of a small group, specialising in safety assessment, who can advise the mainstream design groups on the objectives to be met and the most suitable methods to employ. This group would also assist in the resolution of interface problems and in ensuring that basic assumptions are checked independently.

Relationship with Suppliers of Equipment

Since the prime contractor for the aircraft, as the potential type certificate holder, normally carries the overall responsibility for the aircraft as a whole, it is necessary for him to obtain from the suppliers of equipment such information as is necessary for him to make a satisfactory safety assessment. He should take contractual steps with the object of ensuring the validity of the data provided and its revision throughout the service life of the aircraft. In some cases, e.g. an automatic-flight system, the prime contractor may sub-contract major parts of the safety assessment to the equipment supplier; in such cases it is essential for the equipment supplier to be fully informed, within the terms of his contract, of the relevant airworthiness objectives and the extent of his responsibility. The equipment supplier should accept that his contribution is subject to surveillance by the prime contractor's safety assessment group.

Contracts should be specific about the methods of control of modifications and the use of alternative materials and processes where these could bear on safety assessment.

A problem that can arise with equipment is that in agreeing to achieve defined failure rates, the supplier may be relying on the observance of overhaul periods and inspections which are impractical or uneconomic. There is need for a realistic understanding in this area.

Relationship Between Partners in a Consortium and With Sub-contractors

Tasks may be sub-divided between partners in a consortium, or between the prime contractor for the aircraft and sub-contractors. In such cases the allocation of overall responsibilities must be established clearly so that all interfaces are covered in relation to safety assessments. The use of a small safety assessment group would appear to be necessary to ensure the consistency of treatment and the proper establishment of airworthiness objectives.

DIFFICULTIES AND PITFALLS

General

The analytical methods described in this book can be effective tools during the design and certification of an aircraft. However, these methods result in a considerable amount of work and organisation, so it is important that the assessment is planned in an efficient and effective manner. It should be an integral part of the overall process, and not just superimposed on design and construction tasks.

While maintaining adequately close overall surveillance, it is important to adopt a fairly flexible approach to the choice of methods applied to particular systems or parts of systems. Often, too much reliance is placed on a Failure Modes and Effects Analysis made in a detailed way on piece parts of the system, while ignoring threats which can arise from outside the system hardware, for example, cascade failures or pilot mismanagement.

Numerical Predictions

Numerical predictions can play a most useful part in safety assessment, if used as a method of confirming engineering judgment and for determining check periods or important tasks. However, if they are found to be at variance with past experience or engineering judgment, they should be treated with caution. Conversely, if numerical assessment shows a system to be unacceptable contrary to engineering judgment then the latter will need thorough re-assessment.

Pitfalls

The following is a reiteration of some of the pitfalls previously discussed in this book.

a. Giving too little attention at the design stage to segregation of systems and the possibility of cascade failures.

b. Relying too much on achieving safety by multiplication of identical systems and neglecting the undermining effects of common-mode failures. In the opinion of the authors, one is unlikely to achieve an overall rate of total failure to operate better than 1 in 10 million hours by the use of similar redundancy, so, where higher standards are needed, consideration should be given to reversionary modes or other forms of dissimilar redundancy.

c. Failing to regard the pilot as an integral part of the system with his own failure modes. With actions such as the operation of switches and levers, where there is a chance of unintended or even irrational selection, it is unlikely that pilots' error rate will be better than one in 10 million hours, even with the best of training. It is, therefore, necessary, unless past experience shows that particular types of mistakes never occur, either to make sure that pilot's mistakes will not lead inevitably to catastrophe, or to inhibit the operation of such levers, etc., in circumstances where their misuse could lead to catastrophe.

d. Not conducting proper consultation between all concerned with achieving the safety objectives. This is particularly so when several Organisations are involved, or with bought-out equipment. The solution is largely a matter of organisation, communication, and the use of consistent and comprehensible methods of presenting and recording assessments.

e. Placing too much reliance on predicted reliability figures without considering the sensitivity of the system to variations in failure rates, or taking the steps necessary to ensure that manufacture and maintenance of the equipment is likely to be consistent with the standards needed.

f. Placing reliance on single load paths.

g. Not having independent cross-checking systems so that it is possible for the designer who may make a mistake to be also responsible for checking, with the risk that he may repeat the error, particularly if it is an error in basic assumptions or logic, or lack of appreciation of the airworthiness objectives.

h. Using theoretical predictions of the effects of failures without confirming them on the bench, simulator, or aircraft, over the range of conditions which may apply.

Post-certification Difficulties

As was discussed in Chapter 10, difficulties in applying safety assessment techniques may arise when the aircraft is in service because the burden then falls largely on the operators. It is therefore most important that the operators be provided with sufficient data to assist them to perform their functions. For example, the nature and depth of flight-crew training under simulated failure conditions, and the despatch capability with unserviceable equipment could be affected quite drastically by the assumptions made in the safety assessment.

Most major operators already operate their own surveillance and recording systems tailored to their current needs. They are unlikely to take kindly to radical changes imposed on them apparently as a consequence of the fact that the original design was subject to a safety assessment. It is important that operators should fully appreciate that safety in service can be critically affected by what may appear, superficially, to be minor changes in operational procedure, maintenance periods, etc. It is, therefore, desirable for the aircraft constructors and Airworthiness Authorities to involve the operators concerned as early as is practicable in the evolution of safety assessments so that they are better able to be used effectively later in service.

FUTURE TRENDS

Designs likely to mature within the next decade will involve even more critical uses of systems, some of which will employ digital techniques to achieve the complex functions envisaged. It is likely that micro-processors will be introduced in many systems. Among the types of systems being designed and developed are:–

a. Structural load relief systems.
b. Flutter prevention systems.
c. 'Fly-by-wire', manoeuvre demand, and high authority auto-stabilisation systems.
d. Multi-function cathode ray display systems for basic flight instrument information and general flight management.
e. Digital engine control systems with high authority.

The application of such advances will offer many benefits such as reduced weight, improved flight-path control, higher reliability, reduced work load, and not least, improved safety. The weight saving aspect is becoming increasingly important in the next generation of aircraft in which fuel economy will be vital.

However, demonstrating the reliability and operational acceptability, and establishing compliance with airworthiness requirements are likely to be formidable tasks. It is clearly desirable, when embarking on the introduction of a radical innovation in design, to form a clear view of what is likely to be involved in establishing safety and operational acceptability. The practicability of 'proving' the system within the available time scale may have an influence on the detailed approach to the design. To avoid the risks of entering an open-ended commitment, early consultation between the aircraft constructor, the Airworthiness Authority, and the potential users, is evidently desirable.

An example of the difficulties of establishing compliance with safety criteria is to be seen in systems which rely on similar redundancy to achieve high integrity. Confidence that common-mode failures cannot degrade the high safety level required may take a great deal of time and effort to secure. The alternative of some form of dissimilar redundancy may, therefore, seem preferable on this account.

There are likely to be substantial gains to be had from the application of digital techniques. Three examples can be quoted. It is possible to visualise warning systems which will establish the correct priority of pilot actions, instead of, as at present, the simultaneous display of several warnings which tend to confuse and lead to incorrect action. The use of self-test and built-in test facilities would alleviate the problems of dormant failures. The use of supervisory techniques within the system might be used to reduce the risk of catastrophe from incorrect operation of secondary controls.

As has probably become apparent in this book, the processes of safety assessment are more readily carried out when hardware failures are under consideration. Taking account of crew errors is more problematic. The same can be said about designing with the object of avoiding crew error. Ways and means to improve reliability of the hardware are usually more evident than finding design solutions to minimise danger from crew error. It is necessary to accept the fact that human behaviour is fundamentally prone to error. Training can, and does much to, reduce the frequency of error, but there always remains a residue of risk. By contrast, there is no natural limit to improvement of hardware. Given time, knowledge, and experience, design can always be improved. On these arguments, significant increase of the already high safety levels is more likely to accrue from system design which offers not only higher reliability in its function, but also reduces the demands on the crew and accommodates safely the occasional human error.